T0351925

From Boom to Bubble

From Boom to Bubble

How Finance Built the New Chicago

RACHEL WEBER

The University of Chicago Press
Chicago and London

Rachel Weber is associate professor in the Urban Planning and Policy Department and a faculty fellow in the Great Cities Institute at the University of Illinois at Chicago.

The University of Chicago Press, Chicago 60637
The University of Chicago Press, Ltd., London
© 2015 by The University of Chicago
All rights reserved. Published 2015.
Printed in the United States of America

24 23 22 21 20 19 18 17 16 15   1 2 3 4 5

ISBN-13: 978-0-226-29448-3 (cloth)
ISBN-13: 978-0-226-29451-3 (e-book)

DOI: 10.7208/chicago/9780226294513.001.0001

Library of Congress Cataloging-in-Publication Data

Weber, Rachel, author.
    From boom to bubble : how finance built the new Chicago / Rachel Weber.
      pages ; cm
    Includes bibliographical references and index.
    ISBN 978-0-226-29448-3 (cloth : alk. paper) — ISBN 978-0-226-29451-3 (ebook)
    1. Urban renewal—Illinois—Chicago.   2. Real estate development—Illinois—
Chicago—Finance—History.   3. City planning—Illinois—Chicago.
4. Creative destruction—Illinois—Chicago.   I. Title.
    HT177.C4W43 2015
    307.3'4160977311—dc23

                                                                        2015004406

Contents

Introduction:
Why We Overbuild

The speed of the cycles of prosperity and desolation is an extraordinary challenge to
historians and prophets.

SAUL BELLOW, "Chicago: The City That Was, The City That Is," 1986[1]

LaSalle Street bisects the Loop, the heart of Chicago's central business district, and has served as the city's financial services artery since the 1920s, when
it was known as the "Wall Street of Chicago." A canyon whose steep walls
are carved out of vintage office buildings and ornamented with symbols of
the modern industrial economy, the street terminates on the south in the
imposing Art Deco Board of Trade building (fig. 0.1). A few blocks north sits
City Hall. Viewed as a whole, the street has come to symbolize the solidity
and durability of the City That Works. Cameos of this classic vista have appeared in movies such as *The Dark Knight*, *The Road to Perdition*, and *The
Untouchables*.

In 2006 the City of Chicago declared the area to be "blighted," noting
that the buildings on LaSalle Street were "obsolete."[2] Tenants, it argued, were
rejecting their outmoded structures for the new emerald-glass and steel skyscrapers springing up just blocks away along the western perimeter of the
Loop. It was true: law firms and corporate headquarters had rushed to relocate to the millions of square feet of new office space built since 1998. The
buildings they vacated had a difficult time staying solvent. Some deteriorated,
others were converted to new uses such as hotels, and a few were demolished.
The wrecking balls and construction machinery signaled yet another round
of what Joseph Schumpeter in 1942 famously termed "creative destruction"—
capitalism's habit of rendering old products obsolete and destroying them so
that it can invent new ones.[3]

In this book I explore three aspects of creative destruction in the built
environment that are underappreciated by scholars and practitioners in the
fields of real estate and urban planning. First, creative destruction has a histo-
ricity; innovation and annihilation tend to occur together in time, and certain

FIGURE 0.1. Chicago Board of Trade building, 2011. Photograph courtesy of Kevin Dickert.

periods experience a disproportionate amount of both. At these times the development industry, financial markets, and public planners all race to erect
new structures and demolish older ones. Spectacular changes are made to
the urban fabric. In the following chapters I take apart these dizzying "building booms" to determine what sets them off, what sustains them, and what
cuts them off. Looking specifically at the case of Chicago during the early
years of the twenty-first century, I argue that developers there acted as they
did not because potential occupants demanded new buildings, but because
financial markets were making more capital available. The new office towers
that pulled tenants from LaSalle Street were financed by investors speculating
on future sale prices, prices inflated by the use of derivatives and other complex financial instruments. Under the circumstances, it made more sense for
them to build, buy, and "flip" buildings than to invest in other kinds of assets.

Second, creative destruction is subject to social and political forces that
influence what Schumpeter called its "perennial gale."[4] In addition to their
drive to maximize profit, capitalists are motivated by professional norms—
such as the pursuit of status and peer imitation. Political incentives and regulations also structure and harmonize their behavior. Individual capitalists
receive guidance in interpreting their markets from intermediaries, such as
brokers, who play a critical role in propagating ideas about the right times to
invest and the right buildings in which to invest. Tenants abandoned their

older premises on LaSalle Street because, like developers and investors, they too operated within status hierarchies that placed a premium on the most modern construction. They too were subsidized, either through concessions offered by the landlords of these new buildings or through the City of Chicago's economic development policies. Moreover, they too depended on the advice of intermediaries whose commission and fee structures rewarded mobility and short-distance relocations. Such professional advice helps industry actors make decisions under uncertainty, but can also lead to collectively irrational results. Construction in downtown Chicago, for example, continued unabated during the 2000s despite the fact that employment growth was flat.

Third, during a boom, the actors engaged in creative destruction often misread demand, miscalculating the nature and amount of new products the market is willing to bear as well as underestimating the strength of attachments to the older products that are destroyed. The pull to abide by dominant professional norms is often so strong that it distracts attention from actual occupant demand and results in substantial oversupply. Asset bubbles occur when the market for space comes unhinged from underlying economic activity and prices bear little resemblance to operating incomes. Once burst, bubbles leave behind an overhang of unused space. Vacancy rates spike, particularly in those marginally older buildings whose tenants have flocked to the new stock. On a broader scale, building and lending to excess can bring down the banking system and capital markets, threatening macroeconomic stability.

In exploring these three aspects of creative destruction, some central questions emerge. What caused Chicago and so many other North American cities to boom during the 2000s? What explains the dramatic and rapid changes in urban form witnessed during construction booms, when some areas are built up and out while others are thinned and de-densified? Why do developers seem to get ahead of themselves during these periods, producing more space than is needed by local businesses and households? And what accounts for the fact that some of the historic buildings in older submarkets, such as LaSalle Street, maintain their status and hold their value while others are torn down to make way for new construction during the boom? Rebuilding is usually erratic and incomplete, leaving in its wake, as construction booms do, an assortment of old and new—shiny glass exoskeletons poking out from the bricks and mortar. How does this happen?

To address these questions and contribute to a richer understanding of these periodic booms, their bubble phases, and subsequent busts, I carefully dismantle the urban development process. This process is a mixture of continuity and change, building and unbuilding—dualities more accurately captured by the term "redevelopment." I document the ways in which a combination of new financial instruments and the routine professional practices of private real estate actors and local governments create incentives for capital mobility, new construction, and periodic overbuilding. These practices not only remake the physical scaffolding of cities but also contribute to the financial crises that imperil economies throughout the world.

I focus on what I believe to be the main causes of overbuilding in the United States during the 2000s:

- The financial "innovations" and regulatory changes that create liquidity in global capital markets and set off a chain reaction of acquisitions and new construction that build up some commercial districts while depleting others of investment
- The practices of real estate brokers, investment advisors, and appraisers, who work with occupants and investors to co-construct demand for new products and shift it away from marginally older buildings
- The policies of city governments that provide financial incentives for new building and remove the detritus left over from previous expansions

To illustrate the interplay among these dynamics, I document the case of Chicago's central business district during the period I call the "Millennial Boom," roughly 1998 through 2008.

Chicago was *the* American industrial metropolis, and then it was *the* postindustrial city whose reinvention as a financial services, real estate, and tourism hub—dubbed "the New Chicago"—may have saved it from the same fate as its less fortunate Rust Belt peers.[5] Land economist Homer Hoyt noted that Chicago had quickly developed a reputation for "constantly replacing old tissues with new ones."[6] The Great Chicago Fire of 1871—which destroyed close to a third of the city—was partly attributed to the jumble of poorly constructed frame buildings created over the prior building boom.[7] Speedily rebuilt, the city experienced booms and busts that were the result not of unplanned disasters, but of economic restructuring, demographic changes, public planning, and financial speculation. Its booms were obvious because when it built, it built a lot. Home-grown innovations in building technologies and business organization (Chicago was home to the first general contractor) allowed the city's engineers, architects, and builders to produce some of the world's first, largest, and most celebrated skyscrapers.

Chicago is a self-conscious and frequently studied metropolis, home to many of the scholars who advanced game-changing theories of urban analysis. It has been the focus of scholarly attempts to understand everything from

neighborhood succession and racial change to urban politics to suburbanization and globalization. Its popularity among social scientists is in part a function of its geography. The city's development on a relatively "flat and feature-
less plain" allowed economic, political, and social factors to dictate its future
form obstructed by few topographical features other than Lake Michigan and
the Chicago River.[8] And even the Chicago River's flow could be reversed once
the political will and financial support was secured.

Like Chicago itself, "the Loop," as the city center is commonly known,
has been the prototype for different theoretical models to describe and explain the complexities of urban development.[9] It is the bull's-eye in the con-
centric ring model developed by the Chicago School of sociology in the 1920s;
the hub for models measuring transportation, land uses, and accessibility; the
command center of models of urban politics; and the nucleus for economic
models of urban density and land values. Although urbanized areas have
become increasingly sprawling and decentered, monocentric city models—
with their assumptions of a single employment concentration—continue to
exert a strong pull on scholars. The changes that occurred in the center of
this most central of cities, therefore, have implications for the study of other
North American cities and for urbanism more generally.

Chicago experienced the Millennial Boom in the same way that many
other cities did. During this period, Chicago's downtown added about 15 percent new square footage to its existing stock of office space, which is around
the average for the largest office markets in the United States during that
time. When considered in concert with vacancy rates, the addition of this
new space suggests that the city also *over*built to an average extent during this
period. The expansion exceeded demand: official vacancy rates approached
20 percent after the construction boom subsided, compared with the 9 percent rates when the boom began. It had observers scratching their heads at
the time: Why would building construction and appreciation persist despite
relatively flat economic growth—the signs of which were obvious as early as
2003? After such a disastrous experience with the 1980s boom (when "see-
through" buildings and developer bankruptcies were a familiar occurrence),
how could the industry have let Chicago get overbuilt again?

In the 2000s, most central cities in the United States added copious
amounts of new commercial construction despite slow- or no-growth situations and relatively stable occupant preferences. Although it outpaced competitors such as New York City and Los Angeles, Chicago's bloat was not as
bad as Phoenix's or Atlanta's. The city serves, therefore, as a good case for
interrogating the causes of "routine" overbuilding.

Methods

The book is based on almost a decade's worth of participant observation and interviews during and after the Millennial Boom. I conducted a series of interviews with approximately eighty industry insiders between 2005 and 2014.[10] These informants included lenders at large commercial banks, asset managers at national real estate investment trusts (REITs), pension fund advisors, building managers, space planners and facilities managers for major corporate tenants, market analysts, attorneys, commercial real estate brokers, appraisers, urban planners, architects, and developers (large and small, public, private, and nonprofit).

Because my professional life straddles the worlds of real estate development and academe, I had access to decision makers and providers of development services nationally and in the Chicago region. I used these contacts to locate additional interview subjects who possessed specialized knowledge about their corner of the industry as well as a broader perspective on short- and long-term trends. I spoke their dialect, read their trade publications, attended their professional association meetings, and participated in their social events. On the other hand, as a social scientist schooled in political economy and the sociologies of both technology and the professions, I maintained the distance necessary to interpret the behavior of my fellow practitioners and situate their belief structures in larger theoretical debates. I asked awkward questions and sought explanations for observations that went beyond economistic or self-serving justifications.

This form of elite ethnography was difficult, as the dominant idiom in real estate is a combination of neoclassical economics and psychology. Informants characterized themselves as decision makers on autopilot, responding to price signals beyond their control. And yet they simultaneously subscribed to a theory of the world that granted them considerable agency as strategic risk takers in an environment that was, according to them, entirely of their own making. Their scripts and narratives, so important to constituting what Pierre Bourdieu calls the professional "field," took on renewed importance starting in 2007, as real estate markets across the world faltered and these same individuals sought to avoid personal blame for the pileup of bad investments.[11]

In order to validate the qualitative data collected through the interviews, I also gathered and analyzed quantitative real estate trend data for the 1990s and 2000s. These data included everything from new office square footage to vacancy rates and sale prices to the volume of commercial mortgages securitized over time. My focus was on downtown commercial (i.e., income-generating) buildings that included offices, retail, and multifamily apartments. Although clear parallels can be found for other forms of development, tracking commercial development and tenant relocations allows the relationships between property, financial capital, and tenant movement to be more easily observed because the flows are larger-scaled, better publicized, and take up and release more space.[12] In addition, the private data sources available for analyzing commercial real estate are considered more accurate than those for other types of real estate.[13] Moreover, the topic of residential development, particularly its relationship to subprime mortgages, has garnered much more scholarly and popular attention.

Finally, I reviewed contemporary and historical texts on the appraisal, brokerage, real estate finance, and urban planning professions to trace the origins of constructs such as "Class A, B, and C" buildings, "obsolescence," "highest and best use," and "institutional-grade" assets. These terms represent the constitutive discourses of the real estate field and are critical market devices, particularly in the determination of a property's value.

Outline of the Book

The first part of the book lays out the problem of overbuilding and its causes. Chapter 1 focuses on what the geographer Michael Conzen called "morphological periods" during which political, financial, and social forces are aligned and oriented toward urban transformation.[14] The dominant paradigm articulated in most real estate and urban economics textbooks describes these booms as smooth and cyclical and explains new construction primarily in terms of demand by space users. I argue to the contrary that real estate development is prone to bursts of hyperactivity and dormancy, the motivations for which are often divorced from the business or economic cycle and specific to their times and places.

I use the work of critical urban theorists and economic sociologists to lay out a novel framework for understanding construction booms. The creation of new inventory, spurred by new financial instruments and helped along by the professions that mediate the markets for space, encourages tenants to become more mobile. Strong supply-side factors have been dominant in several, but not all, of the major construction booms of the twentieth century; some booms—like the suburban development explosion of the 1960s— were driven largely by population and employment growth and changes in locational and architectural preferences. In contrast to both supply- and demand-side perspectives, however, I adopt an agent-centered approach to understanding these booms and the surplus of unused space left in their wakes. I paint faces on the professionals responsible for making investment decisions and examine the institutional incentives that encourage them to choose new construction over modernization.

The next two chapters focus on the economic agents whose decisions influence the timing and magnitude of creative destruction in the built environment. Chapter 2 examines the relationship between developers and financial markets. Large-scale developers radically reimagine space and possess the ability to create and extract value from opportunities that few others see. City building depends on their ability to buy and sell land, assemble teams of specialists, execute a vision, and enhance the exchange and use value of property. But developers require the complicity of financial institutions and actors that do not just sanction the projects that developers bring them, but also influence their timing and spatial form. Banks, equity investors, and bondholders influence the signals that help developers decide when it is the right time to build, what they should build, and where they should build it.

Speculative bubbles are the main cause of overproduction. Most readers are familiar with the subprime crisis and meltdown in residential lending that occurred in the wake of the residential overbuilding of the 2000s. In commercial real estate, a comparable trend occurred around the same time. The easy availability of cheap credit and equity, fueled by the demand for new real estate-backed securities, enabled developers of office towers, shopping malls, and multifamily dwellings to move forward with ambitious projects—even in the absence of actual occupants. This temporary easing in credit standards also enabled tenants to borrow more to upgrade to more expensive space. A self-reinforcing spiral of high leverage and speculative price appreciation limited the potential for actors to bring values back in line with operating incomes.

I introduce the players who make up the contemporary system of real estate finance by following what I call a building's "financial life cycle," from construction through the securitization of its debt. Regulatory liberalization, access to new financial instruments, and a surplus of available global capital encouraged these banks and investment funds to underwrite the creation and acquisition of hundreds of millions of square feet of commercial stock that inflated property values and sustained the Millennial Boom. Some of these same actors had been involved in the previous construction boom of the 1980s, but others from that era went bankrupt or were swallowed up by larger firms. The Millennial Boom spawned a new breed of full-service real estate firms and global developers whose financial partners had access to capital through the sale of novel private debt instruments, such as commercial mortgage-backed securities (CMBSs), and public ones, such as tax increment financing (TIF).[15] By 2008 real estate comprised almost 13 percent of the U.S. gross domestic product, the highest share since 1955.[16]

Although my focus is on the specific political and economic context of the early 2000s and I do not seek to advance a timeless theory of overbuilding with immutable laws of motion, I point to a deeply engrained bias toward new construction that is equally evident in prior construction booms. In chapter 3 I identify the local practices that draw global capital to new assets. Local real estate professionals provide market data—lease expirations, market values, rental rates, and absorption figures—to the development and financial actors described in chapter 2 that inform their decisions about whether to build new products and contribute to the growing inventory of buildings. Local actors also reproduce certain practices and beliefs that make modern structures more legible, and therefore more valuable, to absentee owners.

Take the case of commercial real estate brokers. On the surface, brokers simply respond to tenant "needs"—for more or less space, differently configured offices, or amenities. These intermediaries, however, actually help to instill particular conventions and aspirations among tenants. When one major law firm decamps for a new trophy building, for instance, brokers encourage others to follow suit in a kind of arms race for the new space. Competing businesses within the same industry are reluctant to occupy offices vacated by other firms, as leasing second-hand space imparts stigma to certain tenants. I discuss how these and other informal social conventions co-constructed by brokers and tenants encourage the development of new buildings instead of the reuse of existing ones—which helps propel the creative destruction process.

City governments also function as critical market makers during construction booms. They welcome speculative building. Hungry for tax revenues and eager to take credit for new trophy towers, municipalities loosen building restrictions, offer subsidies and infrastructure, and promote new submarkets. Overbuilding then forces cities to intervene at the bottom of food chain, using tax dollars to reposition or demolish buildings vacated by relocating tenants. Cities have historically condemned, sold, and converted older buildings at below-market values to reduce the supply of "blighted" building stock. Dampening supply in this way eliminates some of the competition for tenants and serves to bolster the property values of new construction. These practices, paradoxically, can create even more incentive to overbuild as markets appear tauter and less imbalanced than they really are.

In part II of the book, I present the results of an in-depth study of new office construction and tenant relocations in Chicago's downtown, woven throughout with boom-time stories of real estate developers and investors. Chapter 4 starts out describing the dramatic changes that took place during the Millennial Boom. The downtown added millions of square feet to the existing stock of office space as well as several thousand new residential units each year. This new development contributed to the diversity of uses there and its growing appeal as a place to work, live, and play. But it also hastened the movement of tenants away from other submarkets, accelerating the decline of slightly older buildings in submarkets such as the Central Loop.

Traditional explanations for this building activity look to a growing demand for real estate from businesses and households that are increasing in either number or size. I weigh this possibility against a range of evidence, from employment and demographic statistics to interview data, and find that development did not accelerate to accommodate population and employment growth (the role of income growth is less clear). Occupant interest in the new buildings stemmed primarily from existing businesses short distances away that were stable or, more commonly, decreasing in size. While some businesses moved to new towers that were technologically superior to their existing premises, few appear to have cut costs or raised revenues as a result of their moves. Empirical analyses of changes in business addresses and of interviews with brokers and developers confirm that the Millennial Boom was only weakly demand-led.

Instead, it was capital supply, assisted by the professional conventions of intermediaries and the pro-growth policies of local governments, that drove the building boom. Chapter 5 reveals how the construction of new inventory in the Loop was financed by regional banks, institutional investors, and global investment firms that were flush from selling loans or trading in novel mortgage-backed securities. Under pressure to find alternative sources of profit and desperate for market share, lenders gravitated not only to subprime mortgages for residential borrowers, but also to construction loans for commercial real estate. Chicago was viewed as a city of reliable, if unexciting, investment returns.

Building acquisitions became frenzied as the boom wore on. Almost 85 percent of the office towers downtown were sold at least once, and even half-vacant buildings sold for record-setting prices. What set off this buying spree? The popularity of derivatives such as CMBS contributed to the uptick in buying, selling, and building as they drew new capital into the local real estate market. Buildings increased in value even as the market appeal of the space and services they provided waned in the face of oversupply.

Chapter 6 reveals how the practices of local real estate actors, including the City of Chicago, contributed to the overbuilding and overvaluation of downtown real estate. In a low-growth situation, demand for new buildings must be constructed through marketing and incentives. Tenants were attracted by the generous concessions offered by building owners aware of the oversupply and convinced by arguments made about the obsolescence of their former premises. Tenants also speculated as cheap credit made it possible to upgrade to more expensive buildings and as other industry professionals fed their desire for status and prestige. These strategies worked: the majority of downtown tenants moved at least once during the boom in what amounted to an expensive game of musical chairs.

The City of Chicago helped pave the way for new development on the fringes of the Loop as it sought to realize its decades-long goal of expanding the central business district outward along its western and southern fronts and turning the commercial core into a 24–7 mix of diverse land uses. Raising funds through tax-backed debt instruments, the city invested in infrastructure and provided developers with incentives to construct new towers on the Loop's perimeter. Planners also used these instruments to subsidize the relocation of high-profile tenants and corporate headquarters from older buildings to more recently constructed ones. Commonplace practices such as zoning bonuses and infrastructure upgrading helped the city improve the market for, and therefore the value of, new buildings in hot submarkets such as the West Loop.

At the end of these chains of moves sat marginally older Class A and B buildings, which slipped down the status hierarchy by virtue of their vacancy, deferred maintenance, and smaller tenants. The losers became candidates for demolition or adaptive reuse. Small-scale efforts to recycle "junked" properties rather than demolish them did gain currency in this period, though with some paradoxical effects. While the buildings themselves were preserved, their former uses were eliminated, and in the process, the very markets that were being oversupplied with new products were tightened.

The actors described in this book function as stewards of the existing built environments, and their professional opinions determine the longevity of these environments. In chapter 7 I broach the subject of salvaging the built legacy of cities not for its own sake, but rather to help contain overbuilding. I lay out the opportunities and costs presented by fast and often excessive physical growth and propose policy solutions that could slow the circuits of capital and alter the professional practices described in earlier chapters. These efforts would force the industry to better calibrate spatial expansions to underlying demand and could inhibit future waves of speculative overbuilding.

I assess the prospects of these efforts gaining traction given the growing integration of financial and property markets described in earlier chapters. As the epilogue makes clear, history repeats itself: only five years after one of the worst financial crises in history, prices started to rise and new office buildings and apartment towers were going up in downtown Chicago. The startling consistency with which the same actors undertake the same behaviors underscores the challenges of restructuring a system that provides employment and earnings to so many.

PART ONE

Real Estate Speculations

The Rhythm of Urban Redevelopment

Here, colossal towers are merely placeholders, temporary arrangements of future debris. New York lives by a philosophy of creative destruction. The only thing permanent about real estate is a measured patch of earth and the column of air above it. The rest is disposable.

JUSTIN DAVIDSON, "The Glass Stampede," 2008

If you had stationed a camera at the perimeter of Manhattan's Midtown, downtown Atlanta, or Chicago's Loop in 1998 and taken photographs of the same vista every few months over a period of about ten years, you would have witnessed the scenes in your viewfinder dramatically shifting shape. A modest one-story retail establishment gets replaced with a tendril of glass extending hundreds of feet into the air. A parking lot gives birth to a high-rise hotel. In New York City between 1993 and 2008, more than 76,000 new buildings were constructed and another 83,000 were radically renovated.[1] Metro Atlanta permitted more residential building units than any region in the nation between 2004 and 2007, some 260,000 units overall.[2] In downtown Chicago, twenty-seven buildings over forty stories tall were erected during the period from 1998 to 2010 (fig. 1.1). Some cities gained variety, identity, and texture, whereas others were sanitized by generic corporate architecture.

What this exercise in time-lapse photography reveals is not whether we should welcome or mourn these metamorphoses (whether the new construction is any "better" than the old), but rather the existence of forces at work that periodically transform the urban built environment.

Cities are vast socio-technical ensembles whose actors, relationships, and flows are housed within some of the largest, most expensive, and most durable products of human labor: buildings. Modifying buildings requires power, through the mobilization of resources, political strength, or technical networks. It requires power in the form of financial capital: the debt and equity necessary to acquire land, demolish existing structures, pay for labor and materials, and fund operations. It requires the political power to shepherd projects through a maze of permits, plans, and entitlements. Modifying buildings requires knowledge whose sources are perceived as expert and social com-

FIGURE 1.1. *Top*, Chicago skyline from the southwest, 1971. Photograph courtesy of SOM. *Bottom*, Chicago skyline from the southwest, 2014. Photograph courtesy of Curtis Witek.

mitments that are regarded as credible by members of these ensembles. So when cities resemble preserves of exotic animals—long-necked cranes and big-mouthed bulldozers kicking up dust—it shows that those possessing the requisite power-capital-legitimacy have been enrolled in the project of moving ideas toward material form and are able to overcome the limiting viscosities of place attachment, history, and the status quo.

How regularly do cities undergo such radical physical transformations? And why do these changes so often result in surpluses of unused space? This chapter explores the pace of spatial change and the tendency for property developers to overshoot demand. It then lays out the explanations for (over) building proffered in the past before presenting a novel one that will guide the remainder of the book.

Theories of Urban Change

To many, landscapes are *always* in a state of kinetic and irrepressible flux. Rem Koolhaas calls this state "delirious."[3] Less enthusiastically, novelist Henry James referred to New York City as a "provisional" city that is "defined by a dreadful chill of change."[4] Any experience of the urban is fleeting, Marshall Berman wrote in *All That Is Solid Melts into Air*, as "perpetual disintegration" is part of "the maelstrom of modern life."[5]

Architects and planning theorists have been drawn to this idea of constant change—and not only for the cynical reason that their own professions stand to benefit from cities that are always growing or being recycled. This sentiment captures the modernist, city-as-machine desire for continuous progress that reveals itself variously in authoritarian planning regimes and small-scale efforts at home improvement.[6] Some see the fast pace of spatial transformation as reflecting a distinctly American cultural value:

> Change is integral to the American way of life. America is never finished, and deliberately so. It is a nation with landscapes in perpetual states of emerging and occluding. American places are basically ephemeral, for not even successful places escape constant tinkering.[7]

Virtual ant colonies of industriousness, cities are witness to millions of mundane modifications to the vernacular architecture every day as residents manipulate and personalize their environments.[8]

While rooted in modernism, the notion of constant change also appeals to the postmodern aesthetic, which embraces the flexibility and instability of structures that appear to be durable and fixed. Geographer David Harvey and others have connected the cultural forms associated with postmodernity (including architectural styles that emphasize bricolage and impermanence) to new regimes of flexible accumulation and just-in-time production.[9] Technological innovations in production, globalization, and advanced differentiation in consumption accelerate the pace at which commodities, including buildings, are produced, junked, and reproduced. The inference is that cities are changing at a faster rate than they used to.

The concept of urban impermanence is titillating, profitable, and even fashionable, but is it accurate? Cities may seem as if they are always under construction, with buildings melting into air, but they also exhibit a relative stability of form. Indeed, a less fashion-forward approach to urban change stresses the continuity and obduracy of these anthropogenic environments. In this view, cities are brittle artifacts, like coral reefs, that reflect the cumulative buildup of investment and planning decisions inherited from previous eras. As sociologist Anique Hommels notes, it becomes increasingly "difficult to radically alter a city's design: once in place, urban structures become fixed, obdurate, securely anchored in their own histories and the histories of surrounding structures."[10] Cities represent massive sunk costs not only in large-scale fixed capital, such as buildings and infrastructure, but also in the economic and political arrangements that have evolved to manage that capital. Hommels's study of three postwar redevelopment projects in the Netherlands reveals how difficult it was to radically reshape large, complex sites. Projects there became embedded in such vast networks of actors and social practices, and linked to so many interdependent systems, that mobilizing support for change was nearly impossible to achieve.

Previous development acts as an impediment to radical change as "fixed characteristics restrict the range of possible solutions."[11] Buildings themselves are spatially entrenched commodities whose qualities (e.g., sunk costs, land use regulations, three-dimensionality) resist frequent mutation and hamper swift adjustment to changing demand. "Path dependencies," self-reinforcing dynamics that exert a powerful inertia, make earlier trajectories of development quite durable and difficult to reverse once structures are in place.[12] New construction and demolition may capture the eye, but these events are exceptional when compared with the millions of properties that do not change at all or change in very subtle and slow ways.

Of course, these two accounts of physical change—one emphasizing how cities are set in continuous motion and the other underscoring their stasis—are selective and intentionally extreme representations. All cities experience growth and shrinkage, destruction and rebuilding, as time continually leaves its imprimatur on the built environment. Urban environments are always developing—"charged with predictions and intentions"—and conveying the passage of time (grass growing in a vacant lot, buildings modernized with new fixtures).[13] During some periods that change is fast-paced, while in others it is as slow as syrup.[14] But we often do not notice this change until some episodic event occurs, such as a new real estate venture or the mass clearance of buildings. Architect Aldo Rossi notes that "destruction and demolition,

expropriation and rapid changes in use as the result of speculation and obsolescence are the most recognizable signs of urban dynamics."[15]

Such episodic changes do not take place all the time; the continuum of
urban time is punctuated by bursts of building activity followed by periods
of relative dormancy. Change in the built environment proceeds through
phases of discontinuous speedup and slowdown that are not random, but
linked to other temporally delimited factors.

THE BUILDING CYCLE

Economists characterize such fluctuations in building as "cyclical" in that
characteristics repeat and recur instead of being isolated or random.[16] Like the
general business cycle, these periods typically have four parts—an expansion,
a slowdown, a downturn, and a recovery—and can be represented as short-
or long-wave oscillations.[17] In a short real estate cycle, individual developers
may get ahead of themselves, but markets are assumed to "clear" as rents and
vacancy rates adjust to new supply until demand equals the stock of space.

Starting with Wesley Mitchell and Homer Hoyt in the 1920s and 1930s, the
detailed tracking of land values in cities like Chicago allowed land economists
to empirically measure the length, frequency, peak, trough, amplitude, and
speed of real estate cycles and to search for explanations for them.[18] When
looking at empirical real estate data retrospectively, as they did, one sees an
obvious smoothness to them: no matter what indicator is being tracked (e.g.,
square footage of new construction, vacancy rates, construction employment, investment performance), it looks as if local markets go through long
periods when they are either steadily increasing or decreasing. As an asset
class, property looks more cyclical than, say, the stock market, in which share
prices are often characterized as a "random walk."[19]

One also notices how the cycles of regional real estate markets across the
United States tend to converge, following roughly the same eight- to twelve-
year durations (which includes both their upswings and downswings).[20] While
there is "no place for a model of 'the total construction cycle'" that would
represent all the patterns equally well, certain property classes and kinds of
submarkets often bunch together.[21] For example, regional office space in the
United States has experienced roughly four major trough-to-trough cycles in
the postwar period: the 1960s (1954–1970), late 1970s (1975–1983), late 1980s
(1985–1991), and 2000s (1998–2008). These periodizations are rough because
cycle observers use different measures to detect turning points and because
the start and end dates vary by location. Moreover, some prefer to see longer

construction cycles as composed of multiple, smaller ones (typically lasting sixteen to twenty quarters).

The notion of a cycle assumes that upswings and downswings are destined to repeat themselves. This representation of time is reassuring because cycles are dependable and certain—like those we observe in the natural world (e.g., circadian rhythms, the earth's rotation). But the cycle metaphor may not adequately capture the causes and effects of construction activity. For example, periods of expansion and decline that occur over time within the same geography are not necessarily similar to one another. For one, they leave behind different levels of loss and irreparable change. As Kevin Lynch poetically notes:

> Men have made magical attempts to . . . pretend that change is also cyclical, to imagine that progressive time is a series of eternal, contrasting repetitions, each arising from the other. The magic warms the spirit with the sense that decline and dissolution are only appearances that resurrection will follow. But the things we love do not in fact come back to us. Whatever our hopes, we know things change.[22]

Moreover, the notion of construction cycles, borrowing as it does from the physical and biological sciences, naturalizes change and gives the impression that the rhythms of city building are removed from social interests, economic incentives, or political pressures. This is a common lapse within economics, a field that has a long history of applying "laws" of the "hard" sciences and natural metaphors to the more mutable phenomena in the social realm.[23] Indeed, the study of real estate cycles seems to derive from wave theory, a branch of physics concerned with the properties of undulating processes.

Neoclassical economists are not the only ones who have tried to fit the laws of nature to the artifacts of social life. Marx also saw the tendency of solid material to decompose as a basic fact of modern life.[24] Even urban planner Kevin Lynch gravitates toward biological metaphors to describe the climactic moments and rhythmic repetition of urban environments: they are like a "heartbeat, breathing, sleeping and waking, hunger, the cycles of sun and moon, the seasons, waves, tides, clocks."[25]

Cities and their buildings are not hardwired to behave like either radio waves (whose rhythms recur) or pieces of fruit (whose eventual decay is inevitable). If we look more to empirical social science than to metaphors borrowed from the physical sciences, the rhythms of change appear less regular and more dependent on specific social, economic, and political interventions. They leave cities transformed by their experiences in ways that complicate direct comparison with either previous transformations or with each other.[26]

Indeed, cyclical theory focuses our attention only on those superficial similarities between historical eras—in vacancy or return rates, for example—while avoiding the difficult question of *what has actually changed and why.*

For example, each of the abovementioned office construction cycles left distinctive marks on the cities that experienced them. The 1980s up cycle, for example, looked very different in Dallas than it did in New York City. Figure 1.2, depicting the square footage of new office construction in very different central cities, shows this variation in terms of the magnitude and timing of each city's growth spurts.[27] Manhattan experienced less office construction in the postwar years than it did between the end of the nineteenth century and the Great Depression. In fact, more tall buildings (over 230 feet) were constructed in New York between 1922 and 1931 than in any other ten-year period before or since, and the bulk of them are still in use today.[28] In contrast, Dallas went through a "petrodollar"-fueled office building frenzy in the 1980s. Atlanta feels like an even newer office market, as it experienced more growth in the 1990s and 2000s than Dallas did.

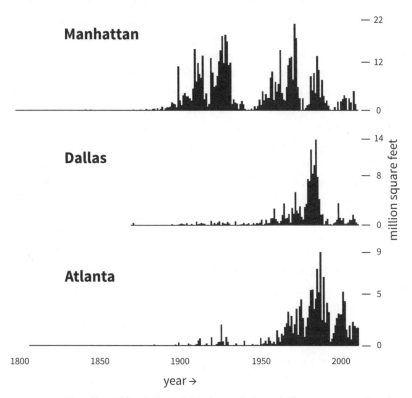

FIGURE 1.2. New office building deliveries in Manhattan, Dallas, and Atlanta, 1800s–2010. Data from CoStar Property Analytics.

Each cycle does not entirely replace, but instead adds to, an inventory of buildings, creating a mishmash of old and new. A crude measure of a cycle's amplitude or degree of change is the amount of new square footage relative to that which existed at its start. For example, between 1978 and 1985, developers erected 36 percent of all the office space ever built in U.S. cities up until that point.[29] In comparison, the cycle that occurred in the first decade of the twenty-first century was half this size: regional office markets in the United States added "only" 18 percent of what existed at the cycle's start in 1998.[30] Buildings produced during each up cycle are bunched into "vintages," the structure and aesthetics of which are superseded by those of the next one. The commercial districts of southern cities such as Dallas and Atlanta feel more contemporary than Manhattan, as they have a larger share of office buildings constructed in the 2000s, even though Manhattan added more square feet of office space during the Millennial Boom.[31]

The fixation on new buildings draws attention away from the losses that occur during construction cycles.[32] New office construction in Manhattan entailed massive amounts of demolition, whereas the development of Dallas's sprawling commercial corridors did not. Younger cities with more developable land and fewer growth controls experience new buildings as a supplement to the total stock. In older, built-out settlements, however, existing buildings are frequently pushed aside for new construction. During the Millennial Boom New York City lost 44,000 buildings,[33] and Los Angeles County lost 44,485 housing units (about 1 percent of Los Angeles's total regional housing stock).[34]

In other cases of redevelopment in older cities, the physical structures remain, but their former uses are eliminated. When an office building is converted into a hotel, the building ceases to be counted as part of the office market and joins the hotel one instead. It is taken "off the market" and put into another one. Some conversions or adaptive reuse projects are ephemeral and informal, so that the switch is not officially recorded. Such is the case when older buildings are resuscitated by artists as temporary sites for work, play, and performance—for example, the Lower Manhattan Cultural Council staged installations and performance pieces in vacant office space throughout the 2000s. When vacancy and absorption rates are calculated, these buildings are still counted as part of the available office inventory (and the art galleries get the boot when paying tenants return).

While giving a sense of the overall magnitude of a city's transformation (that can be compared with previous cycles or at one point in time across cities), aggregate measures do not reveal the configuration or location of the new space *within* a city. Looking at Atlanta's rate of new construction, we do not know if the net square footage added there was the result of twenty new fifty-story skyscrapers or one hundred lower-density buildings. Such differences in form have vastly dissimilar effects on the streetscape, the existing building stock (e.g., the footprint of the buildings demolished to make way for the new construction), and the intangible feeling of a city in the throes or aftermath of a transition. Nor do such measures convey the extent to which, within a single city, the change is concentrated in particular neighborhoods while others are dormant or made worse off by the new buildings. When newly constructed buildings lure away existing tenants in a slow-growing or stagnant economy, the submarkets in which vacancies concentrate will have to sit out a cycle or two. Urban regeneration and ruin are often related; this phenomenon is captured by the term "uneven development."[35] Variegated landscapes allow for the possibility of future transformation as change begets dereliction which begets revitalization.

WHY CITIES BOOM, BUBBLE, AND BUST: WHAT PROPELS CYCLES?

Scholars have focused on the underlying logics that propel this change in regular and easily observed ways. Their explanations for the periodicity of physical change diverge depending on whether they view real property as a commodity valued primarily for its use (to provide shelter, accommodate business tenants) or for its ability to be exchanged.[36] Real estate embodies this duality, which corresponds to the two markets that property serves: that of the occupier and the investor.

The most widely accepted understanding of building cycles, which derives from the discipline of urban economics, privileges the use value of property, which in turn depends on that property's ability to match occupant needs and preferences. This paradigm forms the foundation of real estate economics and is the starting point for most neoclassical models of property markets.[37] Change is driven by the current and potential users of space who seek to maximize their utility; in other words, cycles are market-driven and demand-derived (derived, that is, from household formation, job growth, and business growth). By this logic, the occupants of commercial buildings—the corporate headquarters and national retailers that are their tenants—make profit-maximizing decisions that they express with their feet and wallets. They seek out more rentable square footage if they are expanding and less if they are shrinking—constrained, of course, by the cost of space and their own budgets. Tenants also seek different kinds or amounts of space if they alter their work organization or the floor space consumed per employee.[38]

In these "demand-side" explanations, construction cycles are driven by the aggregation of individual occupants' profit-maximizing decisions to move or expand in competitive markets. The suppliers of space exist to respond to levels of occupant activity; in other words, they are the passive recipients of market dynamics and will build more or differently according to the changing preferences and needs of the intended occupants. The various levels of government, or any kinds of collectives for that matter (e.g., unions), rarely make an appearance in demand-side accounts of urban development. Building is considered a healthy and self-governing process that responds to consumer needs. Any attempt to influence property markets—for example, through local government's limiting growth or channeling it to particular areas—is thought to distort the decision-making calculus of private agents.

Developers are tipped off to changes in demand through price signals. Increases in rent and sale prices announce to them an opportunity to capitalize on unfulfilled occupant interest by bringing more buildings to market. Rent theory, a type of demand-side explanation, predicts that developers respond to rising rents by producing more until the new space is absorbed and the market is pulled back to equilibrium.[39] Economists have demonstrated that supply (particularly office space) is highly responsive to changes in prices.[40]

If real estate is simply derived demand, then spatial change should closely mimic trends in population, net household formation, and employment. In other words, changes in the downtown skyline should tightly align with the general business cycle and follow the general ups and downs of the economy. Increased construction activity may also occur as technological or organizational changes (e.g., replacing humans with computers) decrease the utility of the existing building stock.

In contrast, "supply-side" perspectives tend to view buildings as vehicles through which property owners and others in the industry make money. Construction occurs when the assets produced are considered a good investment—for either an individual homeowner or a Wall Street bank. The profitability of construction as an investment strategy will depend not just on what individual occupants want, but also on broader institutional dynamics: the legalities of property ownership, the cost of capital, the regulatory landscape, and the political power structures of cities seeking the most intensive and fiscally rewarding use of space.[41] As a result, the supply-side approach is more focused on the behavior of local governments and other forms of collective action.

Real estate investment produces an alienable commodity whose association with its location—its spatial embeddedness—makes it scarce. In other words, owners acquire a monopoly on that location and the legal right to capture any increase in the value of the land or improvements there.[42] Owners will finance new construction if returns from rents and future sales are sufficient to pay off the initial development costs and also meet the return requirements and time horizons of creditors and investors. In other cases no new investment is made as land and buildings are merely traded back and forth in order to realize gains from holding, buying, and selling at the right time. Prices are driven not by "the balance between supply and demand but by competitive bidding on a fixed resource by investors who assume that the future price will be greater than the present one."[43]

Investors (or, more generally, "capital") have incentives to keep their funds (also, confusingly enough, "capital") moving fluidly and quickly through the economy. If particular buildings or places present limited opportunities for profit, investors instinctively exit or "switch" their funds. Building on a foundation of Marxist economics, David Harvey argues that the "continuous transformations" of building and unbuilding are a spatial manifestation of the needs of capital accumulation.[44] Capitalists realize more surplus profit if they can decrease the time it takes for capital to be produced, circulate, and turn over. In other words, there are generalized pressures for speedup for both the development of new assets and the acquisition of existing ones, particularly if capitalists can extract additional value out of them (e.g., by raising rents). When capital circulation speeds up faster than new assets can be produced, asset prices can spike and oversupply can occur.

Advocates of this production-oriented approach tend to favor the Keynesian language of "booms," "bubbles," and "busts" over that of cycles. This choice implies that the distinct conditions for each episode are more important than the superficial features that connect them.[45] As metaphors, cycles have incremental beginnings and gently sloping conclusions, whereas booms, bubbles, and busts experience more sudden ruptures and jagged transitions.[46] I too adopt the boom-bubble-bust metaphor to capture the idiosyncratic and irregular nature of these periods of urban redevelopment.

The Conundrum of Overbuilding

These two paradigms would predict that building booms are either tightly aligned with upswings in the economic cycle or dangerously detached from the workings of the so-called productive economy.[47] In most cases they are neither. Instead, the market for space spends much of its time in a state of partial disequilibrium as construction is loosely connected to employment and population growth. Building contractors often refuse to be nudged into action by the creation of new jobs, and development often continues despite weakening in the rental and owner-occupant markets, resulting in what we call "overbuilding."

The case of Houston in the 1980s is especially graphic. The rise in oil prices created a surplus of petrodollars, opening the financial floodgates for new real estate development in the Southwest. The city's inventory of office space went from 70 million square feet in 1981 to 120 million in 1986, an increase of 73 percent.[48] Meanwhile, average occupancy rates dropped from 94 percent to under 70 percent during the same period. Although many of the new buildings remained empty for years, it was the buildings pre-dating the 1980s boom that fared the worst. The owners of the new office towers raided the tenants of the older buildings in order to fill the additional space. Most of the older buildings fell into severe disrepair, and the excess inventory imposed costs on the city through property devaluation and the need for additional public services, such as board-ups and demolition. After 1987, no office buildings were constructed in Houston for eleven years.

During and in the immediate aftermath of the Millennial Boom, over-building again became a commonly diagnosed malady. Observers fretted over development projects halted midstream and empty new subdivisions. Photographers documented overgrown lawns, vegetal swimming pools, and squatters camped out in empty homes.[49] The phenomena of "ghost estates" following Ireland's housing bubble and "phantom malls" in China's newly built cities were widely reported.[50]

Despite our ability to know it when we see it, "overbuilding" is still an awkward and contested term used to capture several different aspects of the demand-supply imbalance. The term is typically used to cast opprobrium: developers are blamed for ruining our environments in selfish pursuit of profit. Measuring overbuilding involves some arbitrary element of comparison (e.g., to long-term growth rates or city averages) and an equally random choice of time period. New buildings that remain poorly leased and half-vacant for years after their construction often find themselves fully occupied during a subsequent expansion. Does that mean they were overbuilt in the first place?

Any attempt to operationalize this fuzzy concept needs to start from the assumption that new supply should hew closely to demand. Any measure of overbuilding, therefore, should account for the magnitude of both new construction activity and vacancies in a given geography. Both elements are critical because otherwise one might mischaracterize places with high vacancies but relatively low rates of new construction as overbuilt. In such cases, high vacancy rates may be caused by decline, due to firm downsizing and closure, rather than overproduction. Moreover, some leasable space should be available to tenants should they wish to move or expand quickly. These so-called natural vacancy rates are region- and sector-specific: a 10 percent office vacancy rate in Washington, DC, may be interpreted as a sign of oversupply, whereas the same rate in Denver may be a limit on future growth.[51] Additionally, high rates of new construction alone do not reflect overbuilding if the new space is entirely absorbed. Cities like New York built extensively during the Millennial Boom but did not get overbuilt because of steadily rising demand for both new and used stock.

To measure office overbuilding during the first decade of the twenty-first century, I multiply new office construction (as a share of the pre-boom square footage) by the difference in vacancy rates between the start and end dates of the boom. I focus on the near term, considering unoccupied office square footage, both new and used, at the end of the construction boom to be in excess.[52] Table 1.1 displays the application of this measure to the fifteen largest regional office markets. The three largest ones, New York City, Los Angeles, and Chicago, fall below the national average. But others, especially

TABLE 1.1. Overbuilding Index, Top Fifteen Regional Office Markets, 1999–2009

Metro	Office inventory, 1999 (million sq. ft.)	Total office deliveries, 2000–2009 (million sq. ft.)	Share new office sq. ft. 2009 (%)	1999 office vacancy (%)	2009 office vacancy (%)	Oversupply index*
Phoenix	102.3	57	56	8.7	20.1	635.19
Atlanta	235.2	63.4	27	8.2	16	210.25
Washington, DC	365.9	104.3	29	5.8	13	205.23
Seattle	146.5	32.8	22	3.5	11.1	170.15
Detroit	169.2	22.6	13	6.2	18.2	160.09
Denver	129	28.3	22	7.4	14.7	159.37
Baltimore	105.38	25.9	25	7.8	12.2	107.97
Chicago	409.4	58.3	14	7.8	15	102.53
Dallas	269.3	64.4	24	13	17	95.66
Houston	231.7	49.2	21	9.4	13.1	78.56
Boston	289.1	38.8	13	5.3	10.7	72.47
Philadelphia	267.4	31.8	12	7.8	11.8	47.55
Pittsburgh	114	12.4	11	8.8	10.7	20.60
New York	1,070.2	42.4	4	6	9.2	12.68
Los Angeles	512.3	55.9	11	11.1	12	9.82
US metro total	3,641.0	669.8	18	9.1	17	145.32

Sources: Data from CoStar Property Analytics and Reis "United States Office, Metro Trend Futures," http://www.reis.com.

Note: Metro areas are Core-Based Statistical Areas as defined by CoStar.

*The index measure was calculated by multiplying the difference in vacancy rates between 2009 and 1999 by the share of new office square footage (of the total stock available in the base year of 1999). This figure was multiplied by 100 so that variation would be more apparent.

those in the Southwest (Phoenix) and Southeast (Atlanta, Washington, DC), stand out for the relative amount of space left fallow.

Because this measure is derived for an entire region, it does not indicate whether the bulk of overbuilding occurred in the central city or in the sub-urbs—or in both. Aggregate measures also do an imperfect job of making visible the internal dynamics between submarkets within one region. New construction often contributes to the most expensive submarkets (Class A office and retail space, luxury condos), and new buildings may attract a suf-ficient number of tenants and buyers. However, the effect of the new units may be to drain the existing inventory of its occupants, forcing those older buildings into a downward spiral of low vacancies and deferred maintenance. In other words, overbuilding is not just a condition that reveals itself when new buildings remain empty; it exists when the addition of new space creates redundancies and raises market-wide vacancy rates.

Scholars on both the demand and supply sides have expended a substan-tial amount of ink trying to explain (and in some cases rationalize) over-building and the chronic lack of synchronization between business and con-struction cycles. For demand-side scholars, the culprit is the delay involved in site acquisition and construction.[53] The fact that it takes, on average, two to three years to deliver a building to market means that construction booms will always be somewhat divorced from demand. Indeed, we often see a two-year delay between vacancy rates decreasing to a low point and average con-struction starts picking up. By the time buildings are constructed and leases are restructured to allow for tenant relocation, vacancy rates have already be-gun to fall.[54] Construction is sometimes said to follow a "hog cycle," drawing a parallel with farmers who raise more pigs when commodity prices are high as they assume such prices are long lasting. Due to delays caused by breeding times, the market is already saturated and prices declining by the time their hogs are ready to slaughter. This process repeats itself cyclically.

For others, the competition for scarce development opportunities keeps markets off-balance.[55] Developers must take control of developable sites and build despite unfavorable current conditions "in order not to be out of the market if it improves."[56] Overbuilding, according to this view, is the by-product of developers building rationally for future, as opposed to present, needs.

Other approaches blame periods of overbuilding on government action, or on the market failures that occasionally prompt government action and distort an otherwise smoothly functioning market. For example, the accel-erated depreciation schedules allowed by Ronald Reagan's 1981 Economic Recovery Tax Act created generous benefits for property owners and were blamed for artificially stimulating construction during the 1980s.[57] Interest rates too are cited as a cause of overbuilding.[58]

In contrast to mainstream economists, critical scholars expect a lack of synchronization between property and business cycles. It makes sense that construction activity intensifies at times when other sectors of the economy are flagging and tenants are shrinking in size because it is at exactly these times that capital "switches" into real estate in search of greater or safer returns.[59] Overbuilding is not an occasional aberration, as mainstream economists would contend, but the spatial articulation of overaccumulation, itself an intrinsic result of the speculative excesses at the root of capitalist production and the credit system.[60] If speculation is the cornerstone of financial capitalism, as Marxists and neo-Keynesians contend, then periodic crises are normal.[61]

When capital extrapolates from past prices and undertakes increasingly risky transactions in hopes of short-term gain, it turns booms into bubbles. Bubbles are sustained by the confidence that additional investors will be drawn to the same slow-moving supply of property and that prices will continue to appreciate.[62] Whereas booms are periods of growth that have foundations in occupant demand, speculative bubbles occur when prices rapidly increase and bear little relationship to underlying sources of value (for stocks, price-earnings ratio; for real estate, building income).[63] Instead, they inflate with hot air.

As long as returns satisfy investors, the suppliers of real estate have no incentive to disinvest. It is only when values decline and investors are wiped out that they pull back from the property sector, seeking returns elsewhere. Real estate bubbles pop. Lenders whose collateral is then worth less than their loans apply more stringent standards for providing credit to applicants. The property sector goes into a period of hibernation until the supply of capital and the mechanisms that switch it are revived and reconstituted.

Overbuilding in Practice: A New Framework

When considered together, the two frameworks that have dominated the study of urban development and overbuilding—neoclassical economics and critical political economy—focus on one of the two drivers of the space market: either its demand or its supply. Neither approach is particularly historical, rarely acknowledging that construction booms occurring at different times may be more or less tied to demand or supply factors. These factors assert themselves with varying intensities in different historical eras.

Moreover, both approaches are analytically debilitated by their lack of attention to the actual actors and institutions that mediate between demand and supply. Mainstream economists who hold to rational expectations assume that actors always make reasoned decisions about development and occupancy based solely on the relative price of space. Their stylized models of markets tend to ignore institutions and collective behaviors that "distort" market rationality. The issue of overbuilding seems to perplex economists partly because it requires them to grapple with the decidedly *political* and *social* nature of real estate markets with all their attendant imperfections (e.g., administrative delays, regulation, poor information, skewed competition, herd mentalities). In contrast, I take a decidedly institutional approach, demonstrating that property markets do not operate on their own, but instead rely heavily on social constructs, market devices, and political interventions to keep them in motion.[64] These conventions alter judgments and can partly explain the apparent irrationality of overbuilding.

Similarly, in the urban political economy, cultural studies, and geography traditions, capital is often characterized as perpetually dynamic and naturally expansionary while the actors and institutions that make capital mobile are deemed irrelevant and unexamined.[65] Urban redevelopment is seen only as a vehicle for the circulation of capital and therefore serves "its" interests with little attention to how differentiated such interests might be.[66] David Harvey notes in passing that capital switching will be accomplished more or less smoothly depending on the efficiency of the mediating institutions.[67] But Harvey and others writing in this vein rarely analyze in any depth the component parts of these institutions. Only tenuously connected to the day-to-day practices of real estate, much of the critical scholarship tends to obscure human behavior, seeing the built environment as the result of "external and autonomous forces with no obvious or necessary involvement of knowledgeable agents."[68] Whether it is consumers or capital investment circuits that drive change and cause overbuilding, these economic agents and structures are depicted as faceless or highly scripted. With institutional analyses of property markets in short supply, "very little is known about the variety of interlocking processes that fuel steep, rapid and highly localized . . . booms."[69]

Focusing on actors does not mean replicating the numerous biographies and autobiographies that provide intimate portraits of the "Men Who Built America." These memoirs of developers reveal their temperaments and attitudes toward risk and inevitably attribute their successes to personal instincts and gut feelings that only their years of experience produce.[70] Nor am I proposing the opposite: a page-turning exposé of malfeasance in the real estate and finance industries.[71] I am less interested in individual stories of brilliant strategy, or of corruption and greed, than in understanding the collective be-havior of developers during booms and bubbles.

In my account, developers are but one node in an industry that comprises highly networked, socially homogeneous, and interdependent professions. The scale, complexity, and public impact of major commercial construction projects require groups of agents contributing multiple kinds of knowledge and legitimacy to the process.[72] Most classical studies of urban development ignore the supporting actors and practices or, at best, relegate them to the sidelines of the drama played out between developers and the state (and, oc-casionally, community-based resistance).[73] But city officials are not the only other critical actors involved in the real estate game; they and the developers that (re)build cities are dependent on both financial capital and sources of professional advice and validation, which they receive from contractors such as appraisers, brokers, and market analysts. These lesser-known professions propagate ideas about the different kinds of value that exist in the built en-vironment, and they guide both public sector planner and private builder toward and away from specific buildings and submarkets.

This book examines the complex professional dynamics that push con-struction booms past their reasonable expiry. Specifically, I document how a combination of financial market innovations and routine professional practices of developers, investors, real estate service providers, and local governments keep markets imbalanced and create incentives for periodic overbuilding. I give voice to agents in these different professions instead of giving agency to abstractions, such as "capitalism" or "markets." Building on the more established supply and demand approaches, I introduce a third framework for understanding the pitch of new construction that periodically transforms cities.

Like political economists, I too assume that the circulation of capital is at the root of most physical changes in the built environment. Yet I show that capital invested in real estate is not innately restless and that cycles are not propelled though their own internal logic and schedule. Instead, capital's movement throughout the built environment must be facilitated and argu-ments for its free passage articulated by actors who possess expert knowledge and stature in the field. Financial and property markets and their boom-bubble-bust cycles must be, in a sense, "performed" through historically and locally specific professional practices. I take my inspiration from institutional economics, science and technology studies, and anthropological and soci-ological accounts of financial markets to understand the engrained world-views, behaviors, and interrelationships of economic agents involved in the production of real estate assets.[74]

Individual actors are organized into professions, "bodies of experts that apply esoteric knowledge to particular cases," that are bound together by credentials, licensing, and ethical codes.[75] The real estate industry comprises myriad professions: brokers, market analysts, attorneys, lenders, building managers, investors, title companies, architects and designers, builders and general contractors, engineers, environmental analysts, appraisers, planners, and surveyors. With the exception of public sector planners, they occupy the "real estate" or "property" sector of private industry.[76] The overall size of this industry has been increasing steadily, despite its ups and downs, since the middle of the twentieth century. In 2008, before the downturn hit, 1.5 million Americans worked in the real estate industry, while 7.2 million worked in construction.[77]

I focus on the subset of professions engaged most directly in knitting together property and financial markets. Situated between developers and both their financiers and tenants is a set of intermediary professions that includes brokers, appraisers, and local planners. By positioning buildings in relation to others, these actors guide investment into specific property classes. Absent the systems of meaning they reproduce, property could not be constructed, sold, or securitized.[78]

These professions have worked hard to disassociate themselves from their nineteenth-century origins as "curbstoners," "land sharks," and "fly-by-nights" with a penchant for swindling and self-dealing.[79] Progressive-era values of service, specialization, and technique led to the development of membership associations and techniques designed to make their advice more neutral and trustworthy. Modern forms of corporate organization and less permeable boundaries between experts and the public allowed these professionals to better navigate between the interests of clients and allegiance to "objective truths" about the markets they were serving and creating.

Despite their professionalization efforts and new calculative rationalities, these intermediaries also play a role in overbuilding the markets in which they operate. They do not do so just by making a few tactical mistakes. Instead, their work, at its core, helps make risky behaviors appear sensible. Constructing buildings with few tenant commitments, acquiring half-empty buildings, purchasing bundled mortgages as securities despite questions about the creditworthiness of the underlying income streams, and making investments in less-tested markets are all behaviors common to construction booms. In contrast to observers who believe that increasing professionalization will reduce tendencies toward speculation and crisis,[80] I show how certain professional behaviors actually enable bubbles.

Routine professional practices contribute to overbuilding by ensuring a consistency of worldview. Professions function as "epistemic communities" with shared beliefs and notions of validity.[81] Despite their biographers' focus on individualized traits, actors in the real estate field adopt common practices through repeated interactions, membership in professional organizations, and dependence on related sources of remuneration. Moreover, social similarity in the occupational makeup of the property sector leads to groupthink across the industry. Others have pointed to the social psychological processes that produce a kind of boom "mentality"—usually one associated with an unfounded optimism whereby actors overestimate how singular their investments are and fail to take into account the inevitable lemming-like response to profitable ones.[82] Rather than focusing on individual behaviors, I highlight the interlinked and mutually dependent professional incentive structures that reward similar kinds of performance and thrive on mimicry. When decisions must be made under great urgency and uncertainty (as is the case with speculation), actors are even more drawn toward standards, norms, and assumptions shared by others in their fields. Their attention lingers on their peers and competitors, not on the broader markets they help construct. Acting in concert is one of the dynamics that produces the smoothness observed in cycles—both the herding to get in as prices rise and to get out as they fall.

Take, for example, the practice common across the real estate industry of associating age and functionality. Specifically, professionals tend to stigmatize structures built during certain historical eras as "obsolete."[83] When they do, they breathe life into a concept with a long and troubled history: writing in the United States in the 1920s and 1930s, the land economists who were the founding fathers of the real estate profession determined that buildings had innate life cycles, with average life expectancies ranging from twenty to thirty years depending on building type. Their early treatises on valuation endowed the concept of obsolescence with a technical legitimacy and social efficacy. Ideas about natural building life spans and narratives of obsolescence and modernity, with their stories about the irreparable damage that time and progress had wrought, have endured and are still used by brokers, city governments, and appraisers, who must weigh in on the very subjective notion of a property's true value. The use of the obsolescence discourse by these professionals "at once poeticize[s] capitalism's chaotic mutability as a natural condition and deploy[s] rational analysis to contain the inherent anarchy of creative destruction."[84]

Not all construction booms depend as heavily on narratives of obsolescence. During the employment- and population-driven upswings that occurred in U.S. cities during the first half of the 1920s and in the suburbs of the 1960s, the pressure to quickly accommodate growth encouraged both new construction and, when that could not keep pace with new demand, building modernization to meet contemporary standards. Moreover, the construct of obsolescence is not hegemonic; it is occasionally contested by historic preservation advocates, environmentalists, smaller landlords and tenants in older buildings and submarkets, and city planners who question not only the dominant interpretations of value but also the constant circulation and placelessness of capital.

Narratives of obsolescence are employed in full force, however, when construction disassociates from growth—as it does during speculative bubbles. When tenant demand for new construction is weak, development professionals must find other compelling rationales to justify the fast-paced acquisition, construction, and lease transactions that animate and remunerate their work. The concept of obsolescence allows capital to release itself from preexisting commitments and move on to more profitable arrangements. Even when economies are not growing, a confidence pervades the profession that tenants will always leave their current premises if a more modern product is available ("if we build it, they will come"). It is easy to see how such thinking can lead developers to overshoot actual tenant demand for new space and result in overbuilding.

I take the position that the real estate industry constructs and perpetuates certain meanings that *make* the very markets they seek to accommodate. Supply helps to influence demand. Examining the supply side in more depth is an antidote to the strong demand-side orientation adopted by the majority of professionals in the real estate field. When questioned, real estate professionals modestly suggest that their job is to accommodate the changing needs of potential building occupants—for more or less space or for more technologically sophisticated buildings. They believe that they are simply passive receptors for market demand, which exists a priori and out of their realm of influence. This deep-seated belief in consumer sovereignty assumes that tenant needs and preferences alone determine when and what their industry builds.

Most demand-side explanations assume that tenants always want the latest building materials, layout, design, and technology because these amenities make them better off. Such assumptions fit with the notion that society is propelled by a progressive impulse toward innovation and positive change, articulated clearly by Schumpeter's notion of creative destruction.[85] Developers, then, are not just building new, but they are also building better—spurred on by tenants' demand for higher-performing space. Technological advances are assumed to prompt replacement rounds of buildings as older ones become obsolete and abruptly fall out of favor.

Yet previous studies on depreciation and my own interviews with tenants, developers, financiers, and service providers revealed occupant needs to be ambiguous and quite pliable.[86] Office tenants are often on the fence about whether relocation will benefit them materially. Some already lease space in buildings that meet their functional requirements or that could do so with a little upgrading. Rarely do tenants commission their own new buildings or provide in-depth input into the design of a new building. Instead, blue-chip developers acquire sites and make plans to construct new towers years *before* talking to tenants: "We tee up all these deals and wait for conditions to be right to move forward," noted one.[87] Buildings start off as sets of specifications that integrate the latest technologies and anticipate the tastes of their intended end users in the abstract (glass curtain walls, maximum usable floor space). And it is important to keep in mind that the end users for a new income-generating building are not only the tenants, but also the investors to whom the developer seeks to sell the asset in the future. Developers then market the building extensively and provide incentives (naming rights, rent concessions, generous improvement budgets) for tenants to relocate there.

Tenants are responsible for addressing their own specific spatial needs once the financing and approvals are in place to move construction forward. During the design phase that follows, tenants hire architects, engineers, and contractors to customize the interior layouts, systems, and finishes. After negotiating who will pay for these improvements, the parties eventually inch toward closure, and the developer completes construction in time for the start date of the lease.[88]

In many ways, the suppliers of new buildings "configure" tenant needs, giving occupants a sense of what is available and desirable. The appeal to tenant preferences as justification for adding new inventory is often an after-the-fact legitimating device that obscures the interest developers have in creating new assets that will generate fees and profits.

That said, I do not mean to imply that tenants have no agency or that they are coerced into making decisions that go against their best interests.[89] Developers and real estate brokers, for example, cannot sell a tenant on a new building that does not conform to what it wants and expects. Supply-side actors are not capable of imposing their own perception of which qualities should matter most to a tenant's business operations.

Moreover, the tenants that occupy Class A office space are shrewd, calculating agents themselves, capable of perceiving differences and responsible for making weighty judgments under great uncertainty. The decision whether to renew one's lease or relocate to a new building nearby is rarely a straightforward one, as few large-scale businesses inhabit workplaces that are crumbling to the ground. The quality of their existing premises and the pace at which it declines vary considerably. Corporate tenants have to balance several interests: to use space more efficiently, adopt the right aesthetic signifiers to convey status and prestige (down to the designer of the furniture in the waiting areas), control their environments over the long term, appeal simultaneously to employees and clients, manage costs, and offer locational amenities such as proximity to train stations and services. It is difficult to generalize about which of these varied interests predominates in each case because each tenant's use of space and priorities are so unique. It is clear, however, that through repeated social and professional interactions with those on the supply side, as well as monetary incentives, many tenants came to hold similar understandings: that new products are superior.

The following example illustrates the difficulty of singling out the motivations for the major tenant moves that, in turn, anchored the new construction of the Millennial Boom. Developers and brokers found many reasons to besmirch office buildings from the 1980s, but they singled out "inefficient floor plans" as the leading cause of their obsolescence. Indeed, during the 2000s many large law and finance firms expressed similar concerns: the physical constraints of the square and octagonal buildings they leased were cutting into productivity and profit. "If you find yourself in a building that has a lot of interior space, you end up having to rent the floor to get the exterior windows and buying a lot of interior space that you then use for paper storage. The result of it is you're spending $40-a-square-foot to store paper. By moving to a differently configured building, we were able to save a lot of dead space," noted law firm Kirkland & Ellis, which in 2009 moved west from Chicago's Aon Building (built in 1974) to a new office tower.[90] Large-block tenants wanted to "stuff more headcount into smaller spaces,"[91] and they could achieve the desired employee densities in newer buildings that were designed with column-less floors and fewer personal offices. Moreover, new buildings tend to devote fewer square feet to common spaces such as lobbies, whose maintenance and operations costs are shared by all of the buildings' tenants.[92]

But other tenants with comparable footprints stayed put in their 1970s and 1980s buildings with "obsolete" floor plans, deciding that renovating them was less expensive and disruptive than moving to new premises. The facilities manager for one Chicago law firm opined, "We could have definitely stayed in our old space and retrofitted it to get away from that 1980s–1990s interior. You can pretty much renovate anything these days, even the exteriors . . . When we looked at the two options side by side—to stay or go—we were talking about pretty marginal improvements and similar costs [going with the option to move]."[93]

Tenants have intrinsic needs, but they are also influenced by the developers, brokers, and designers whose livelihoods depend on filling the new buildings and who vigorously promote the improvements offered by new construction. Tenants also measure their own success in relation to the behavior of their peers facing similar decisions. Such complex interactions make it impossible to identify which set of actors drives the trend toward more open office formats.

I propose that demand for new buildings is "coproduced" through tenants' ongoing interactions with developers, brokers, and others who move fluidly between occupant and investment markets.[94] Less a theory than an idiom, coproduction involves a process of mutual adaptation between what suppliers propose and what consumers want. It implies that production and consumption are more closely related than those on either end of the paradigmatic spectrum imagine: supply and demand factors are present in all booms and are intertwined.[95] For example, tenants stimulate developers to innovate (to build open floor plans), and developers expect tenants to consume features that they may not want or need but have become industry standards (to use open floor plans). Moreover, tenants use the end products they occupy in ways unforeseen by developers, and, with the buildings and their income streams, developers play investment games about which tenants know very little. Coproduction offers a sensible middle ground between supply- and demand-side perspectives, accepting that new construction is always influenced by some combination of occupant need and investor interests.

The following two chapters explore in more depth two underappreciated and interrelated causes of periodic overbuilding on the supply side: financial markets and the professional practices in which they are embedded. Chapter 2 describes the basic financial foundations for commercial real estate development, without which it is not possible to understand how cities boom. Chapter 3 singles out some of the key meanings real estate intermediaries attach to buildings that help justify the creation of new structures during boom times. These justifications and meanings are highly consequential, as they not only shape design trajectories and tenant preferences, but also determine the extent to which overbuilding becomes a problem.

Fast Money Builds the Speculative City

How can you tell when the real estate market is overbuilt?
When lenders stop lending.
TRADITIONAL REAL ESTATE JOKE

At the start of the twentieth century, the ability to own or trade in real estate for investment purposes was highly restricted in the United States. In Illinois, for example, it was illegal for corporations to own real property beyond "that needed in the transaction of their corporate business."[1]

At the start of the twenty-first century, however, real estate investment is ubiquitous. Any corporation, individual, or household with disposable income can purchase property outright, regardless of whether it intends to occupy it. A car dealer in Benton, Texas, can own a share of the repayment stream for debt held on a mall in suburban Philadelphia, while a teacher in Oregon's public school system can purchase a slice of a Chicago office tower through her pension fund. New investment vehicles, professional practices, and legal-institutional reforms have transformed property into a fungible commodity, one bought and sold in increasingly liquid markets. The widespread popularity of real estate has galvanized the financial sector, which controls the unprecedented volume of capital coursing through the concrete.

Scholars refer to both the increasing *size* of financial markets and institutions and the increasing *role* of finance in nonfinancial sectors as "financialization," arguing that these processes have occurred generally across the economy over time.[2] The central organizing principle of contemporary capitalism is its dependence on the short-term flows of global finance, the "infrastructure of the infrastructure."[3] In this phase of capitalist development, profits are made more through financial engineering than through commodity production and trade, as they were during earlier eras. Evidence of this shift includes the size of the financial sector: the value of all financial assets in the United States grew from four times the gross domestic product in 1980 to ten times the GDP in 2007.[4] Financialization can also be detected in the more subtle changes in the orientation of nonfinancial sectors, such as real estate, health care, or public policy, as global investors acquire new powers to "discipline" economic behavior.[5]

The fact that cities take on the characteristics of their financial sponsors is nothing new; medieval cities like Florence and imperial cities like Istanbul embodied the desires and status hierarchies of their kings and dynastic families. Today the car dealer, the teacher, and their financial fiduciaries have also left their imprimatur on cities. The elevated skylines of financial districts such as Wall Street and the City of London, as well as disinvestment in other city neighborhoods, speak to specific instances of the financial sector's significance as the gatekeeper to credit and the enabler, therefore, of construction and rehabilitation. Financial actors can determine *when* cities grow (i.e., the timing of construction booms) as well as *how* and *where* they grow (i.e., how much new inventory is produced during these periods and the types of buildings and locations that will dominate during booms).

And yet the mainstream real estate literature tends to treat finance as a follower, not a leader, in the realm of property development. Finance exists to validate the impulse to build, which is driven by the interplay between developers and occupants. In the corporate (auto)biographies of real estate giants such as James Rouse and Donald Trump, it is developers who are the lone visionaries in whose hands the keys to the city can be found.[6] Their appetites for risk pit developers *against* financiers, who are stereotyped in these narratives as small-minded and risk averse. Bankers and investors enter into the picture only when developers seek "backing" to make their grandiose plans a reality.

Conversely, critical scholarship tends to restrict the agency of developers by treating them as tools of capital and regarding real estate as a passive outlet for the surpluses generated in other sectors.[7] Developers are driven by the innate "needs" of capital to switch between different circuits of accumulation and root out the highest profits.

In contrast to both literatures, this chapter examines the ways in which financial actors interact with and influence the built environment. These actors mobilize one another in shifting sets of ventures that create new assets, convert existing properties into more standardized and legible commodities, and keep markets liquid and fast-moving. This chapter first guides readers step by step through the financial phases of a typical income-generating building. It then discusses the regulatory interventions and market practices that increased the volume, reach, and speed of capital seeking investment opportunities in real estate during the booms of the 1980s and 2000s. As real estate professionals channel more capital into the property sector, developers construct buildings that are less of a response to occupant needs than a reaction to prices proffered for debt and equity in capital markets. It is not surprising, therefore, that during periods of credit expansion, supply outpaces demand.

The Financial Life Cycle of Buildings

The impetus for new commercial construction is elusive. Occasionally corporations commission developers for "build-to-suit" projects they intend to own or rent.[8] In most instances, however, developers must interpret aggregate data, the advice of experts, and individual instinct to imagine demand for their products.

Absent a single client with deep pockets, developers turn to the financial sector to build buildings and develop land. Real estate development is a highly leveraged business, more so than other sectors of the economy.[9] Over time, property owners have decreased their reliance on savings and sought financing through increasingly complex channels. They now shop around for large amounts of "other people's money"—equity and debt financing—to underwrite their construction projects and the continued operations of their buildings. Developers require not just good ideas or occupant demand, but an abundance of capital, which means they are "resource dependent"—and therefore subject to pressure from those agents that control the necessary resources.[10]

Developers' reliance on external sources of finance stems partly from the fact that real estate is too expensive, complex, and long-lived to be paid for in cash up front. Their dependence also derives from the organizational structure of their firms. The field of development comprises a range of differently sized corporations; most are loosely conjoined operations that alter their legal and organizational form to abet their current projects. The larger and better-known firms—for example, Tishman Speyer, Hines, Trammel Crow, Trump, and The Related Companies—are diversified in scope, international in coverage, and in some cases publicly traded.[11] They are vertically integrated and operate lending, acquisition and disposition, valuation, site selection, construction, brokerage, and management arms.[12] For example, Tishman Speyer not only develops real estate but also manages a $34 billion portfolio of 72 million square feet of commercial space—the rough equivalent of all the office space in Houston and Los Angeles combined—spanning four continents.[13]

Despite the size and omnipresence of a few full-service global companies, the most common type of development firm operates in smaller regional markets.[14] These firms have interchangeable names that few outside of their localized markets would recognize, such as MR Properties LLC (Chicago), Madison Park Financial Corporation (Oakland), and Hand Properties (Atlanta). Small developers are provisional species with the ability to grow or shrink rapidly. They maintain their flexibility by keeping a low profile and low overhead, making up-front commitments to contractors and service providers outside the firm (e.g., architects, builders) that come due as each phase of development is completed.

The assets that result from the labor of developers are often removed from the conditions surrounding their birth. Once erected, the buildings are adopted by a series of new investors who view property as they would stocks and bonds: as an income-generating asset. But commercial property is not a natural-born investment vehicle; neither the production nor the transfer of real estate is conducive to the fast circulation of capital. In fact, property markets are regarded as some of the most illiquid ones available to investors, in that products are not standardized and cannot be "sold continuously at a price that everyone in the market can know."[15] The quality of a high-rise office building in San Jose is not public knowledge, nor is it easily fathomed— particularly if the investor is located in Memphis. Despite the proliferation of private subscription data services, accurate information about the condition of specific assets is asymmetrically distributed and heavily guarded by local actors.

A complex web of local spatial relations imbues property with value. Even if a building is fully tenanted, its value fluctuates depending on what neighboring owners do, whether local governments invest in nearby amenities and infrastructure, and whether the demographic composition of the area changes. These local idiosyncrasies "translate into financial risks that can impede liquidity and discourage the development of cross-border exchange and global flows."[16] Such risks are easily hidden in the convoluted spreadsheet models and accounting tricks common to this sector.[17]

Moreover, money invested in the built environment (as opposed to other assets) is less divisible because it is immobilized for long periods. Such illiquidity makes assets particularly susceptible to devalorization and depreciation.[18] Combined, these characteristics make real estate a quintessentially "opaque" commodity that requires both highly localized knowledge to grasp and an extended amount of time to trade.[19]

How is such an opaque and socially embedded asset financed? The typical income-generating building experiences a "life cycle" of financial flows, from its first construction loan through its securitization. I conceive of buildings not just as places and spaces, but as sets of events defined by the movement of capital into and out of them. Applying this framework allows us to track capital's progression from its most material, slow-moving embodiment (i.e., a construction project) to its most ethereal and expeditiously traded form (i.e., the building's mortgage pooled, sold, and resold as a security to investors).[20]

CONSTRUCTION

The volume of commitments they juggle and the lumpiness of cash flows prevent even the largest of developers from using their own retained earnings to fund current projects. Profits from previous projects acquired through management and development fees, capital gains, and building revenues may provide them with a portion of the equity needed to cover predevelopment expenses. Developers can also access equity from sources such as insurance companies, real estate investment syndicates and trusts (REITs), and investment funds, but these funds are expensive and rarely sufficient to cover the bulk of construction costs. If they borrow money, however, developers can deduct their mortgage interest—along with depreciation and investment losses—from their taxable income, creating an incentive for leverage. As such, most developers secure debt to build.[21]

Because it requires such high degrees of specialized knowledge and numerous contractual safeguards, the majority of lenders lack the stomach for construction, or what is known as "project," financing. If a real estate project is finished and occupied within a reasonable time frame, however, returns can be significantly higher for construction financing than for other forms of lending.[22] High up-front fees, the ability to get in on the ground floor, and the availability of customized insurance products also make construction loans attractive to lenders.

The main sources of construction finance are commercial banks. The industry is highly segmented, and typically the larger a construction project is, the larger the lending institution. During the Millennial Boom the chief underwriters for the new Class A office buildings and condo towers included U.S.-based "money center" banks such as Wells Fargo, Wachovia, Bank of America, and JPMorgan Chase; regional banks such as LaSalle (Chicago) and PNC Bank (Pittsburgh); and international banks such as Deutsche Bank and the Royal Bank of Scotland.[23]

In addition to private banks and investors, local governments (including cities, counties, states, and special authorities) also extend credit and equity to private development projects.[24] In the name of "economic development," the public sector may pay for the costs of land acquisition, infrastructure, and construction that developers otherwise would have to shoulder themselves. It may do so through "soft second" (below market rate) loans, tax credit equity, and the allocation of tax dollars through vehicles such as tax increment financing, a tool that allows future property taxes to be committed to current development projects.

Private and public sources provide different forms of debt to developers (fig. 2.1). They may provide pre-construction loans, which advance working capital to borrowers engaged in new ventures. These monies are some of the first invested in the project and are not secured by it. Similarly risky, land development loans are used to purchase and prepare raw land for construction. These loans may be repaid from the sale of improved lots to other builders or bundled into a construction loan.

While these two sources of capital are optional, most developers must secure at least one commercial construction loan. Construction loans are short-term debt that have one- to three-year terms and carry higher interest rates than longer-term commercial mortgages. They are "drawn down" as the developer compensates members of the development team for work completed. Interest accrues on the rapidly growing principal.

Construction lenders have short-term time horizons; they want to get their money into and out of the project as quickly as possible. But erecting a new building can take several years; legal regulations (such as zoning and building codes), convoluted contracts, and the uncertainties plaguing

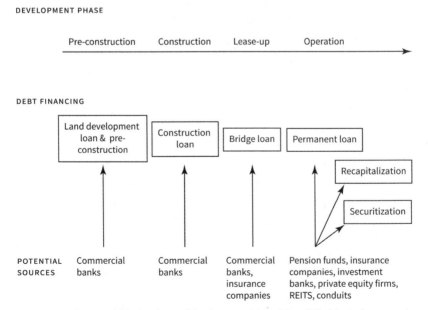

FIGURE 2.1. Sources of debt for phases of development. Adapted from "The 'classical' construction finance structure," MIT Open CourseWare, David Geltner, Real Estate Finance and Investment class.

the construction process (complications around site control, impediments to clear title, material shortages, and labor delays) slow production. Banks' collateral is worthless unless the project is completed, and there may be no market or future owner for the project when it is. As such, lenders require a market analysis, a report written by a third party that measures potential demand for the proposed space. They also scrutinize the future revenue-generating potential of the borrowers and projects they are considering financing by applying underwriting guidelines that dictate acceptable standards for pre-leasing and pre-sales, permanent financing, valuation, and even the terms for hiring contractors. Risk management strategies such as these are intended to guard against cost overruns and the possibility that banks will have to take title to a half-finished shopping mall because construction stalled or no tenants located there.

OCCUPANCY AND OPERATION

One of the dynamic tensions within real estate development is the need to reconcile the risk preferences and time horizons of different financing sources. In a typical construction project, this is accomplished by having at least two separate instruments—one short-term, one long-term—to cover the construction period and the separate operating period of the building. Construction lenders typically require a precommitted, "permanent" loan to "take out," or pay back, the construction loan before their loans even close.[25] This practice helps protect banks from the possibility of providing credit to a project that may not make it to the point where it can be leased, refinanced, sold, or securitized.

In contrast to construction loans, permanent mortgages are secured by the built and occupied project. They are "permanent" in theory because they may be the last loans that the project carries (although in practice, most are refinanced). Loans for commercial projects often amortize only a small portion of the principal and therefore require a large balloon payment and re-capitalization through a new loan or sale at maturity. Providers of permanent loans scrutinize projects based on their income-generating potential and potential risks. Specifically, projects need to be able to quickly generate sufficient cash flows from rents and sales to cover debt service payments and operating expenses.

Developers may work with mortgage brokers and bankers, who originate loans but fund them from short-term lines of credit extended by other banks. Or they may borrow directly from commercial banks, which include everything from money center banks to smaller state-chartered banks to regional savings and loans. The array of banks holding commercial mortgage debt is diverse, in contrast to the residential mortgage market, in which a few, very large banks have historically advanced the majority of funds.[26] During the 2000s money center banks held the largest portfolios of commercial real estate (CRE) loans, but regional banks also loaned to smaller and riskier projects.[27]

Federal and state efforts to deregulate banking in the aftermath of the economic crisis of the 1970s allowed banks to offer a wider array of loan products and lowered the accounting standards they used to report their financial conditions.[28] The elimination of interest rate ceilings on deposits, for example, allowed banks to set rates and price their own loan products. Relaxation of geographic restrictions that governed the flow of credit across national and state borders promoted a more extensive circulation of capital.[29] Those banks that remained standing after the waves of mergers and acquisitions following deregulation used a variety of new interbank debt instruments to put idle reserves into circulation, moving massive sums of money throughout capital markets in response to local demand conditions.[30]

Deregulation forced banks to compete with one another for yields across a national and increasingly international playing field.[31] Banks also faced competition from capital markets and from nonbank lenders such as institutional investors, private equity firms, and trusts, which ventured into development finance as a way to shore up profit rates in their core operations.[32] Although typically more active in the equity and recapitalization markets, these institutions originate loans at higher interest rates than banks and also provide equity for real estate deals. For example, mortgage real estate investment trusts (REITs) make loans secured by real estate, and since the 1920s insurance companies have created subsidiaries to originate debt, lending out the funds they hold for their clients.

RECAPITALIZATION

Most real estate projects are owned by a single-purpose entity or limited liability corporation (LLC), which obtains financing for that particular project. Ownership structures are likely to change several times over the life span of a building, as few developers build with the intention of holding the property in perpetuity. Instead, most "flip," or sell their products to new owners, shortly after construction and stabilized occupation. While holding periods for real estate are longer than for stocks and bonds, they have been declining over time.[33] Shifting money around frequently from one asset or fund to another has become a sign of financial acumen and, if the timing is right, a way to increase short-term returns from the capital gains generated by the sale of buildings.

Building owners typically pay only the interest on loans for periods of five to ten years and refinance the principal payment when the loan comes due. Even before the loan matures or refinancing occurs, the capital structure of the project may be transformed by a sale or recapitalization. Specialized "recap" firms—typically private equity firms, investment funds, or REITs— borrow money to purchase controlling stakes in assets in hopes of reaping profits. The recap process involves exchanging one form of financing for another, such as more debt for less equity, or replacing more expensive debt with a cheaper kind. It may also involve physically transforming buildings through renovation so that the new owners can raise rents and attract a different tenant mix.

The acquisition of Manhattan's Stuyvesant Town by a partnership between developer Tishman Speyer and asset manager BlackRock illustrates this process. A sprawling complex of fifty-six red-brick apartment buildings, Stuyvesant Town was built and owned by Metropolitan Life Insurance Company to house returning war veterans in the 1940s. In the 2000s the partnership borrowed $4.4 billion from sources such as the government of Singapore and raised equity from investors such as the California Public Employees' Retirement System and the Church of England to acquire the property.[34] It paid $5.4 billion in 2006—the highest price for a single residential property in the United States—with the intention of converting ("deregulating") the rent-stabilized apartments to market rents. The financial crisis that began in 2007 caused the partnership to default on its loans; it sold the property back to its lenders in lieu of foreclosing. Tishman Speyer's adept use of other people's money allowed it to lose "only" $56 million on the failed deal, as opposed to its investors and lenders, who took a $2.4 billion hit.[35]

That MetLife held onto Stuyvesant Town for more than sixty years is unusual. More commonly, prospective buyers start to circle new buildings as soon as they are erected. Publicly traded REITs, for example, take ownership stakes in shopping centers, office buildings, apartments, and hotels from developers when the new projects start generating cash flow. They are added to the REIT's portfolio of properties, shares of which are sold on stock exchanges. Through such vehicles, investors are able to buy and sell interests in diverse real estate properties expeditiously without either the expertise or the commitment of time and money needed for direct property ownership. Such low barriers to entry partly explain the growing popularity of REITs: in 1972, the market capitalization of the entire REIT industry was $1.5 billion; in 2006, it peaked at $438 billion, with 183 active funds.[36]

Institutional investors such as pension funds and insurance companies also recapitalize real estate to hedge inflation and diversify their portfolios. For example, insurer Aetna was active in commercial real estate in the 1980s, and Prudential was a key player in the late 1990s and early 2000s.[37] Traditionally known as risk averse because of their legal obligations to policyholders and pensioners, these investors are interested in sole or joint ownership of well-leased properties and typically hold onto them for five to seven years.[38] Like other financial sources, however, they loosen their underwriting standards during booms to maintain returns, increase market share, and meet their obligations.[39]

Wall Street investment firms recapitalize the projects of prominent developers, not only backing their acquisition efforts with large mortgages, but often taking an equity stake in the deals to pump up offers in heated bidding wars. For example, Lehman Brothers' commercial real estate strategy— which caused both its "swanlike transformation from a second-tier bond trading shop into a full-service investment bank" and its ultimate demise— involved forging close ties with global developers such as Tishman Speyer.[40] Lehman bankrolled the acquisition of trophy buildings in New York City, such as the Chrysler, Woolworth, MetLife, and Seagram Buildings, as well as entire portfolios owned by other firms. It put its own cash into these deals as well as funds from the pension funds, educational endowments, and other institutional investors whose investments it was managing. Fearful of the kind of exposure such arrangements entailed, Lehman would sell its equity stakes as quickly as possible once deals had closed. But when the markets started collapsing in 2007, the firm found itself with billions of dollars of overpriced commercial real estate that it could not unload.

Private equity firms (often called "buyout firms") also make short-term loans and take equity positions in real estate projects in exchange for high returns. While this capital is more expensive, it often comes with more lax underwriting standards. Private equity firms maximize their returns through leverage; the less equity they must put into an acquisition deal, the greater the eventual return. These firms often swoop in before a new project is fully leased if they can improve operating income or make a quick sale, often to another private equity firm. In 2010 the top three commercial landlords in New York City were all publicly traded equity funds.[41]

SECURITIZATION

Banks and nonbank lenders historically held most of the loans they originated in their own investment portfolios. Since the early 1990s, however, loan originators have sold these mortgages to third parties, such as mortgage bankers and investment firms. The third parties then pool and package the loans into commercial mortgage-backed securities (CMBSs). In other words, banks have become conduits for larger loan pools and the front line of a complex chain of securitization.

Securitization is the practice of separating income streams from their underlying assets and selling off the rights to them.[42] Income streams are pooled with comparable products and repackaged as bonds for investors in capital markets. As with REITs, investors effectively purchase a share of the pooled income stream, but unlike REITs, these securities are built on debt service payments rather than direct ownership of the asset. Mortgages are converted into new assets through techniques that disaggregate and continually reassign ownership in order to allow for more and faster-paced exchanges. In exchange for this liquidity, bondholders must contend with volatile yields, as the prices of these assets are more sensitive to fluctuations in the economy than the appraised values of the underlying real property.[43]

Investment firms simultaneously serve as originators, underwriters, and distributors of these securities. They have been playing these roles since the 1920s, when merchant banks and bond houses issued securities backed by commercial buildings whose construction they financed.[44] Investment banks sell mortgages either to institutional investors or to other divisions within their own firms, transferring the risks of the asset side of their balance sheets to other parties. Before the financial crisis in 2007, for example, over 70 percent all outstanding CMBSs were held by institutional investors such as insurance companies and mutual funds.[45]

Why do lenders, big and small, engage in such practices? Securitization allows participating banks to rid themselves of the liabilities, risk, and inflexibility associated with holding longer-term debt as mortgages are swept off the originator's balance sheet. It also allows banks to bypass "capital adequacy" requirements, regulations put in place to ensure that banks have a buffer of equity and savings to guard against potential losses.[46] But perhaps the most immediate motivation is that mortgage originators, as well as their brokers and the investment firms reselling the mortgages, collect substantial fees for each service provided and each transaction completed. In 2007 alone these transactions provided banks with over $530 million in fees.[47] Selling debt-backed securities, therefore, pumps excess cash into the earlier stages of a building's financial life cycle.[48] In other words, it allows lenders to raise short-term funds to lend out to developers looking to build and investors looking to buy.

Once erected, new buildings start generating the rents and debt service payments that are the primary cash flows supplied to capital markets. Investment firms bundle the mortgage payments of one building along with those from hundreds of other buildings into one pool owned by a trustee. The trustee collects the cash flows and distributes them in waterfall fashion to holders of differently priced tranches (i.e., segments of the pool that have different risk-return profiles). Less risky slices might go to institutional investors while the lower-rated ones are sold to hedge funds. In the process, the different income streams associated with one office tower may be "owned" titularly by several distinct entities.

Credit rating agencies, such as Moody's and Standard & Poor's, are responsible for reviewing the underlying loans to determine the risks associated with each tranche. They assign grades to structured credit products that both reflect and determine demand for these products. And, like the appraisers described in the following chapter, they are paid by the very firms whose products they rate.[49]

Although securitization represents the apex of a building's financial life cycle, neither the capital invested in a building nor the bricks-and-mortar structure itself "dies" at the end of cycle. Instead, securitization completes the financial life cycle of a building by turning something as idiosyncratic and long-term as real estate into fungible commercial paper that trades quickly in public markets. A private market transaction between an originating lender and a borrower is converted into a public market transaction between bond investors, banks, and their intermediaries. Financial markets release property from its intrinsic embeddedness. When mortgages are pooled with others to become a mortgage bond, that bond becomes more alienable than a building, "an abstract object but one that is capable of ready conversion into an understandable exchange value."[50] Stripping assets of their distinctiveness creates uniformities and facilitates their quick sale. Through this process property inches closer to a "pure" financial commodity, while finance appears more locally embodied and "real."[51]

The effects of securitization reverberate back to the prior stages in a building's financial life cycle. Securitization broadens the base of possible capital sources and expands the volume of funds available to refinance existing buildings or, importantly, erect new ones. When asset-backed bonds sell for escalating prices in capital markets, it sends a strong signal to loan officers to make more loans. Indeed, the interlocking gears of these financial relationships require a constant supply of new assets to keep them in motion. Noted one developer, "There are a lot of different ways to make money off of things that get built, but not a lot of ways to make money off of things that don't."[52] Real estate development and capital markets are mutually constituted; they build each other up and, just as quickly, bring each other down.

Switching to Drive: Financing the Construction Cycle

Forget location, location, location. It's now timing, timing, timing.
National Real Estate Investor, October 20, 2010[53]

The financial sector regulates the schedule of real estate production, influencing both the timing of construction booms and the physical texture of those cities that experience them. "The availability and price of finance capital—in the form of mortgages, commercial loans, and direct investments—heavily condition where, when, and how much commercial and residential property will be constructed" in any one place and time.[54] The duration and volatility of the boom-bubble-bust sequence depend on the amount of capital banks and investment firms provide and the terms on which they provide it. One real estate professional put it succinctly: "There is a belief that bankers finance what builders build. In fact, the opposite is true. Money is the *sine qua non*. If you have money, you will build."[55]

BOOM-TIME BEHAVIORS

The financial sector does not provide credit or investment capital to all places and building types at all times. David Harvey, the most prominent analyst of the phenomenon of "capital switching," argues that markets for the development and acquisition of property (the "secondary circuit") provide an outlet for surplus reserves of capital fleeing the "primary circuit" of basic commodity production when returns there are relatively low.[56] Although the primary and secondary circuits are now more integrated than Harvey imagined, if capital switching theories are valid, we should see returns from real estate investment lagging those, say, in manufacturing or technology as fund managers rebalance portfolios when stock prices in those sectors decline.[57] The potentially zero-sum relationship between the primary and secondary circuits helps explain why real estate investment cycles are often divorced from business cycles.

And yet even when capital is tempted to switch to the property sector because of low returns elsewhere, it cannot do so on its own. Capital circulation must be engineered by the institutions and actors that mediate between the supply of capital and the demand for it. Loan officers at banks, pension advisors, and investment portfolio managers at insurance companies, as well as brokers, appraisers, and planners, facilitate the movement of capital between financial and property markets. These intermediaries employ practices, such as underwriting standards, that guide the flow and determine the pace of capital moving into real estate.

When underwriting standards are restrictive, professional relationships fraught, or returns uncertain, capital is more likely to linger in other sectors. Building activity wanes. Cities are more brittle and resistant to overt transformation. A handful of experienced, well-capitalized developers may obtain credit for projects that demonstrate clear market demand. Construction lenders see projects that are the best credit risks during periods of restricted capital access: those in excellent locations, with occupant commitments for the entirety of the leasable area, and on the verge of completion. If they are not interested in building, developers may seek funding to acquire cut-rate property left over from previous booms, or they may reallocate resources to their brokerage and asset management operations to wait out the capital drought.

Once the choicest projects have been financed and are generating cash flow at appropriate levels, financial markets may shift to favor borrowers— particularly if there is sufficient competition for assets. Trying to replicate their early successes, lenders and equity investors seek more outlets for their capital. If the economy is growing and tenant demand keeps up with new development, a healthy construction boom results.

However, these same high yields can draw more savings and surpluses into the property sector. In the event of a capital bounty, financiers loosen their underwriting standards: they lower debt service requirements, allow higher loan-to-cost ratios, and provide more nonrecourse and interest-only loans.[58] As the supply of capital grows and chases fewer worthy projects, financiers become supplicants, and developers-owners pass on more risk to them. Developers rush to bring new construction online to keep ahead of their competitors.[59]

The relationship between supply and market demand then starts to unravel. With banks requiring fewer pre-sale commitments and guarantees, developers are free to propose projects that are more speculative; that is, they can construct new buildings without commitments from tenants. During such times developers are just as prone to discounting the market analyses they have commissioned as their market analysts are to inflating projections of near-term demand. Developers, investors, and lenders put their faith in the ability of new buildings to pull tenants from existing ones. In the process, they antagonize other investors and landlords and lower overall occupancy rates and rents. If they have limited financial interests in the existing stock (for example, if they are new to the market), suppliers of new buildings have no incentive to stop this process of "cannibalization."[60]

Upward price movement encourages more investors to enter the acquisitions market and speeds the pace of transactional activity.[61] Developers are able to sell their newly constructed buildings for prices in excess of their costs of production plus a "normal" profit. A spate of new construction activity can push markets toward a new, higher pricing structure, rather than helping to equilibrate a previous one, as appraisers weight their estimates of value toward recent comparable sales—no matter how unrealistic the sale prices.[62] As paper values inflate, debt-to-equity ratios decrease, and both lender and borrower appear to be more financially solvent than they really are.[63] Speculative bubbles are recognizable when rapid appreciation occurs despite no concomitant increase in building income.[64]

Witnessing rapid escalation in property values, fledgling development teams enter the market, producing a kind of bandwagon effect. With no track record and little capital of their own, they borrow whatever is made available to them. Once financial gatekeepers have lowered their standards, even these untested entrants are treated as credible borrowers. In the words of one broker, "People [are] tripping over each other to lend the money, so guys just [take] it."[65] Lacking the intuitive feel for market saturation that seasoned developers may possess, they are prone to more reckless building that flouts occupant demand.[66]

When developers can easily sell their products and lenders can easily sell their debt, moral hazards can result. Able to make an easy exit, they ignore a building's revenue-generating potential and take on a "gambler's mentality and reckless disregard for risk."[67] That risk, after all, becomes someone else's problem once the building's ownership is transferred.

Likewise, active secondary debt markets encourage lenders to decrease loan quality; conduit banks can make loans without having to absorb the risk of default or illiquidity. By shifting loans off their balance sheets, lenders are free to continue issuing debt and collecting fees. Eventually they stop "securitizing [mortgages] in order to lend" and begin "lending in order to securitize."[68] The reallocation of risk shifts the entire market in the direction of increasingly marginal projects and "explosive cycles of speculative investment . . . until markets are dangerously overbuilt and falsely inflated equity values can no longer conceal the deterioration of earnings."[69]

Developers are bolder in liquid markets; they build not because of a pressing need for space on the part of occupants, but because of the availability of capital to build, an acquisitions market ready to pay top dollar for whatever they can erect quickly, and capital markets willing to competitively price their debt for sale to distant investors. The trick, experienced professionals know, is timing: to "ride the bubble" for as long as they can before the fever pitch of transactions cools down.

A TRUNCATED HISTORY OF CAPITAL SWITCHING

Capital supply and construction booms are connected in history, theory, and practice. Relatively low rates of return in alternative asset classes (such as corporate equities and technology stocks) combined with low interest rates and new securities have led to reallocations of investment portfolios favoring real estate.[70] Throughout the twentieth century real estate booms and bubbles have immediately followed the peaks of business cycles and continued several years after their tail ends.[71] Investors have used real estate assets strategically to hedge risks from other investments and take advantage of pricing differences between them (i.e., arbitrage). Introductions of new instruments and aids to liquidity have historically triggered building jags; for example, the issuance of mortgage-backed bonds "financed the construction of most of the U.S. skyscrapers in the 1920s and led to overbuilding and then widespread vacancies."[72] The "market acceptability" of these risky instruments is short, generally five years, after which new ones take their place.[73]

Capital markets possess the power to convert booms supported by demand into speculative bubbles, causing prices to spike rapidly and building cycles to skip a down cycle.[74] While most short-term cycles play themselves out every seven to ten years, financial bubbles are more unpredictable.[75] Keynes foresaw such hyperextensions of the business cycle, arguing that speculative behavior amplifies the cycles' ups and downs and prevents them from adjusting to the oversupply of new assets.[76] These tendencies are exacerbated in real estate cycles because, compared with corporate equities, property is less liquid, and real estate bubbles take a longer time to deflate than stock market ones. Several of the twentieth century's recessions were delayed for a few years because of the trading and collateral value of property.[77]

The demand-driven expansions that characterize booms are distinct from the finance-led real estate bubbles that extend them. Most of the regional growth spurts of the 1960s, for example, particularly the rapid suburbanization of metropolitan regions, were a response to the postwar buildup of household wealth and the expansion of the middle class—aided by generous federal housing and transportation policies.[78] The postwar boom was in fact several smaller "boomlets" that did not leave a substantial overhang of unused space in their wakes. When increased market demand (from population and income growth or a dramatic shift in preferences and building quality) is the primary catalyst for building, disequilibria between supply and demand can be ameliorated by lowering prices. That is, an oversupply of space declines in value due to the dampening effect of rising vacancy rates, and construction is halted or slowed.

   In contrast, finance-fueled bubbles tend to end suddenly and badly. Mass speculation in commercial real estate was tied not only to the global financial crisis of 2007–2008, but also to the Great Depression, the savings and loan crisis in the United States,[79] Japan's lost decade,[80] the Nordic crisis of the 1980s,[81] and the Southeast Asian financial crisis in the 1990s.[82]

   The remainder of this chapter describes two recent commercial real estate bubbles in the United States—in the latter halves of the 1980s and 2000s—which both resulted in substantial overbuilding. Although they involved different financing mechanisms and locations, patterns emerge with regard to their characteristics and drivers: mass interest in debt-backed instruments that recapitalized banks and lowered underwriting standards, tax laws and policy changes that encouraged real estate investment, and declining profitability in other economic sectors that prompted investors to park their wealth in the built environment (table 2.1).

The 1980s

Perhaps the most obvious example of capital switching in recent U.S. history occurred in the late 1970s, when both the financial and property sectors went into a prolonged ascent while many other sectors of the economy stagnated. The dismantling of the Bretton Woods international monetary management system in 1973 and the recession of 1973–1974 are generally regarded as turning points in a long wave of economic expansion that had been in progress since the end of World War II.[83] Sharp declines in corporate output and profit rates at this time intensified the search for returns as well as struggles between labor and capital over the distribution of those returns.

   This period also marks the beginning of what observers refer to as "the socialization of finance," or "the ownership society," as ordinary households invested their savings in securities, were switched to "defined contribution" pension plans, and, accordingly, developed a vested interest in financial market performance.[84] Laying the groundwork for a new phase of public sector financialization, state and local governments experimented with increasingly risky investment schemes to pay operating expenses, plug budget holes, and push obligations further into the future. They added new financial instruments such as variable rate debt, interest rate swaps, auction bonds, and derivatives to their own debt portfolios—often with disastrous effects (e.g., the case of Orange County, California, and its foray into swaps in 1991).[85]

   During this time commercial real estate investment was dominated by partnerships seeking tax shelters for money made elsewhere. Also active were

TABLE 2.1. Comparison of 1980s and 2000s Commercial Real Estate Booms

	1982–1991	1998–2008
Main actors	Private and publicly held real estate developers Tax-oriented limited partnerships Commercial banks Savings institutions Institutional investors Insurance companies	Private and publicly held real estate developers Investment banks Commercial banks Institutional investors REITs Municipal governments
Regulatory changes	Employee Retirement Security Act of 1974 created incentives for institutional investment in real estate Economic Recovery Tax Act of 1981 shortened commercial depreciation period and reduced the top marginal personal tax rate Depository Institutions Deregulation and Monetary Control Act of 1980 and the Garn–St. Germain Depository Institutions Act of 1982 deregulated banking industry Tax Reform Act of 1986 eliminated accelerated depreciation	Reigle-Neal Interstate Banking and Branching Act of 1994 removed restrictions on opening bank branches across state lines Gramm-Leach-Bliley Act of 1999 effectively repealed Glass-Steagall's separation between investment and banking Commodity Futures Modernization Act of 2000 deregulated transactions involving derivatives
Motivations	Growth in service industries, decline in goods production Growing integration between real estate and capital markets Baby boomers and women increased size of white-collar labor force Shifting locational preferences: to south and from city to suburbs	Commercial mortgage securitization lowers cost of debt financing Increased employment in financial and business services Rezoning and public financing provide incentives for downtown development Shifting locational preferences: to south and from suburbs to city
Locations	Sunbelt metros (Dallas, Houston) Western metros (Oklahoma City, Tulsa, Denver, Orange County, Sacramento) Northeastern metros (Boston)	Sunbelt metros (Phoenix, Las Vegas) Southeastern metros (Atlanta, suburban Virginia) Western cities (Seattle) Downtowns of traditional urban centers (Chicago, New York City)

Sources: Adapted from C. Alan Garner, "Is commercial real estate reliving the 1980s and early 1990s?," *Economic Review* 93, no. 3 (2008): 89–115; Alan Rabinowitz, *The Real Estate Gamble* (New York: American Management Association, 1980).

banks, which shifted greater proportions of their loan portfolios into real estate to fend off competition from nonbanks and adjust to tighter margins. Commercial banks' exposure to real estate (including construction and permanent loans as well as home mortgages) hovered around 30 percent of their total asset values in the period between 1960 and 1985.[86]

Globalization made more capital available worldwide, and an increasingly integrated financial architecture allowed surpluses from oil (petrodollars generated in the Middle East and the United States) and trade (generated in Germany and Japan) to move freely to investments abroad.[87] Foreign contributions to the U.S. real estate sector increased dramatically during the 1980s.[88]

The task of guiding capital through turbulent markets and switching it between investment circuits fell to professionals in the finance, insurance, and real estate (FIRE) sector. The private equity firms, the commercial and investment banks, and the business services that supported them expanded in size. Growth in the number of office workers exceeded 4 percent annually in the late 1970s and continued throughout most of the 1980s.[89] In a recursive manner, those industries facilitating capital switching into real estate experienced the employment growth necessary to stimulate and sustain the early part of this boom. Developers initially put up new buildings to accommodate expanding FIRE firms, which sought buildings with more space, state-of-the-art communications technologies, and flexible floor plans.[90]

Commercial construction occurred primarily in Sunbelt cities and their suburbs, pushing at the peripheries of metropolitan areas.[91] Suburban office parks sprouted up across the country, and shopping malls dotted the sprawling landscapes. Downtown markets in cities across the Northeast and Midwest were also affected: Washington, DC, Manhattan, Boston, and Chicago all expanded the available square footage with bold new skyscrapers.

Federal policy facilitated the switch. Banking deregulation allowed savings and loans to shift funds made from selling residential mortgages into commercial construction and acquisition in the hope of getting a better return in the face of double-digit inflation.[92] The tax treatment of real estate changed with the Economic Recovery Tax Act (ERTA) of 1981, which reduced depreciation schedules, allowing investors to write down the full value of their commercial property investments over a fifteen-year period (shortened from a forty-year one) despite actual anticipated life spans of a much longer duration. In addition, ERTA introduced the Accelerated Cost Recovery System, which "front-loaded" depreciation, providing proportionately greater write-offs in the early years than did the traditional straight-line depreciation method, but smaller ones in later years. New buildings became attractive tax shelters because the dollar value of the deduction was higher when the majority of an asset's depreciable base value was intact. This and the fact that properties could be resold and depreciated several times ("churned") encouraged investors to switch their surpluses from corporate stocks and bonds into bricks and mortar.[93]

The bubble portion of this boom is often attributed to the passage of ERTA.[94] Shortly after its passage, the constant dollar value of commercial construction increased 50 percent in just two years (1983–1985),[95] and the value of commercial construction as a share of GDP jumped to 2.0 percent (from 1.3 percent in 1979).[96] Almost half of all office space that existed in 1990 in the United States was built during this period.[97]

Unfortunately, this was also the era of the so-called see-through building.[98] By the end of the decade, the national downtown office vacancy rate was close to 20 percent, and rents and property values were sinking rapidly.[99] The recession first hit the "oil patch" states of Texas, Louisiana, and Oklahoma, moving north to New England and eventually to the West Coast.[100] Were it not for Japanese investors and publicly traded REITs, the appeal of ERTA in 1986 might have single-handedly brought down the property markets. As it stood, it took several more years for construction to subside. Overseas investment was cut off in 1991 when crises hit Europe and Asia and exchange rate fluctuations scared off investors. Although owners received tax write-offs for their failed investments, economic losses from the oversupply of offices in the United States during the 1980s amounted to at least $130 billion.[101] The commercial real estate sector went into hibernation for most of the 1990s, only to reemerge at the turn of the century.

The 2000s

After the bust of the early 1990s, Wall Street banks and financial advisors devoted resources to data gathering, forecasting, and investment analysis in the property sector. The resulting intelligence, observers believed, combined with the financial monitoring provided by increased securitization, would dampen future real estate cycles and make them less volatile. Experts were uniformly subdued as building activity began picking up in the late 1990s; one noted that "the current market cycle is demonstrating how well the feedback loop of information available today is helping to slow construction in most markets in the U.S."[102]

Capital sources began to ramp up slowly over the 1990s. Trade surpluses (this time from China and the Middle East) and subsequent crises in East Asia and Russia made investors punch-drunk on one hand and skittish on the other. Traditional safe harbors such as U.S. Treasury bonds repelled investors with their unusually low interest rates. The technology bubble burst in the latter part of the 1990s, and the stock market was twitchy. Uncertainty about oil prices and foreign wars kept capital searching for cover and close to the ground.

Real estate, the most grounded of all asset classes, became the favored option as property values depreciated more slowly than the price of technology

and corporate stocks did. Private equity firms, REITs, and institutional investors switched domestic funds into the construction and acquisition of commercial real estate. They also shepherded sovereign wealth funds there: foreign entities owned approximately 15 percent of the total equities outstanding
in the United States.[103]

Mortgage-backed securities coupled with a global capital glut helped ig-
nite the building frenzy of the 2000s. Investors looked to the active secondary
markets in equity and debt-backed securities, initially purchasing shares of
REITs during the 1990s and then bonds secured by pooled mortgages in the
2000s. Commercial banks held the majority of mortgages for commercial
real estate during this period, but CMBS syndicators outpaced life insurance
companies and government-sponsored entities as the second largest holders
of such debt (fig. 2.2).

CMBSs had emerged out of the Resolution Trust Corporation (RTC),
which the federal government created in 1989 to turn the commercial real
estate holdings of insolvent savings and loans into derivatives—one of many
instances of the government "capitalizing on crisis."[104] Market acceptance
of CMBSs grew during the 2000s, after a few initial issuances in 1992 and
the global liquidity crunch following the Russian financial crisis of 1998. The

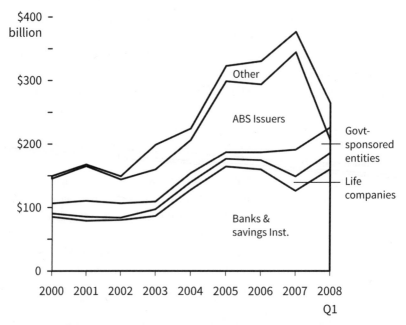

FIGURE 2.2. Commercial mortgage flows, 2000–2008. Data from the Federal Reserve, Flow of Funds
Accounts of the United States.

Gramm-Leach-Bliley Act of 1999 knocked down the walls separating banking from investment that the New Deal's Glass-Steagall Act had erected. After 1999 full-service investment firms could originate mortgages in addition to providing equity, channel clients' savings into new investment vehicles, and provide advisory services. In the late 1980s and 1990s, large investment banks such as Drexel Burnham Lambert issued these securities, and in the 2000s a different crop—Wachovia, Morgan Stanley, and JPMorgan Chase—took the lead (table 2.2).

Whereas less than 2 percent of commercial mortgages were securitized in the 1990s, over one-third of the commercial real estate loans originated between 1998 and 2008 were structured as CMBSs.[105] Issuances of CMBSs peaked in 2007, when more than $913 billion of the $3.3 trillion in outstanding loans was pooled and converted into these instruments. Adding yet another layer of complexity, CMBS bonds were bundled into even larger pools and converted into collateralized debt obligations, or CDOs. The CDO market outpaced traditional buyers of subordinated commercial real estate debt, and like the CMBS market, it peaked in early 2007.[106] Most of these mortgages were used to buy existing office buildings and retail centers; multifamily apartment buildings and hotels were less popular property types.

Securitized debt drew more capital into commercial real estate, shutting out insurance companies and publicly traded REITs by 2006.[107] Nonsecuritizing lenders were also left in the dust. The CMBS market tended to cherry-pick the bigger deals, so small and medium-sized regional banks barreled their way into increasingly risky sectors: notably commercial real estate and

TABLE 2.2. Top Ten CMBS Issuers, United States, 2006 and 2007

	2006				2007			
	Ranking	Issuance (billion $)	Number of deals	Market share (%)	Ranking	Issuance (billion $)	Number of deals	Market share (%)
Wachovia	1	27.1	13	13.2	2	30.7	10	13.3
JPMorgan Chase	2	25.9	12	12.6	3	26.5	9	11.5
Morgan Stanley	3	25.0	21	12.1	1	34.2	24	14.9
Credit Suisse	4	19.3	11	9.4	7	17.7	9	7.7
Banc of America	5	19.0	16	9.2	5	19.4	10	8.4
Merrill Lynch	6	17.4	9	8.5	4	20.6	10	9
Lehman Brothers	7	17.2	10	8.4	6	18.9	7	8.2
Deutsche Bank	8	14.6	11	7.1	9	12.2	7	5.3
Citigroup	9	11.4	10	5.6	10	11.5	7	5
Bear Stearns	10	11.3	14	5.5	8	14.7	14	6.4

Source: Data from Commercial Mortgage Alert, "Top CMBS Bookrunners," https://www.cmalert.com/rankings .pl?Q=68.

eventually subprime household mortgages.[108] Repeating the mistakes of the 1980s, banks again expanded into commercial property; the total volume of commercial mortgages held by banks more than tripled between 1998 and 2008. Compared with the 1980s, when real estate made up about 30 percent of banks' asset value, this share averaged 54 percent in the 2000s. For many smaller community banks, this share rose over 300 percent.[109] Indeed, it was the small and medium-sized banks—the middle-market segment—that cornered about 70 percent of the commercial real estate loan market.[110] Construction loans were some of the fastest-growing mortgage products, increasing by about 10 percent annually. In 2006 construction loans as a share of total assets of commercial banks and savings institutions reached the same level as they did in 1985.[111]

Local governments fueled these flames by extending credit to private development projects using non-guaranteed debt such as revenue bonds.[112] After holding steady and even declining for most of the 1990s, outstanding state and local debt in the United States increased by 62 percent between 1997 and 2008 to $2.58 trillion (of which approximately $1.56 trillion was local debt).[113] With these funds, cities allocated subsidies to private developers and built ambitious infrastructure projects themselves.

In many cities, demand for the new space could not keep up. Employment growth in the FIRE sector flattened in the 2000s.[114] Moreover, the rest of the economy grew at a slower pace than it did in the 1980s, with observers noting a "jobless recovery" after the 2001 recession.[115] What Harvey calls the "demand problem" was "bridged" by more debt financing for both developers and occupants as rising indebtedness hid repressed wages and slow economic growth.[116]

Developers did not seem to notice as they contributed more stock to locations with preexisting gluts. While the boom of the 2000s is primarily known as a residential boom, particularly in the "sand states" of Florida, California, Arizona, and Nevada, commercial lending and construction also continued unabated during this period.[117] Developers built, confident that even if they lacked tenants to occupy the new space, appreciation and acquisition activity would continue unabated. Private sector (nongovernmental) nonresidential construction put in place in the United States reached $410 billion in 2008.[118]

Although developers added about half the *volume* of new commercial construction that they did in the 1980s, asset *values* increased at a faster rate during the Millennial Boom.[119] From Miami to San Diego to Chicago, too much money chasing a finite set of deals caused commercial properties to inflate in value. Prices came to bear scant relation to the material condition of the underlying assets and the income those assets were capable of generating.

The commercial sector experienced the lowest capitalization rates in history. The "cap rate" is the ratio of a building's income to its sale price, so it measures the degree to which acquisition markets are unhinged from occupant ones. Average cap rates for the office sector dropped from over 9 percent in 1992 to an unprecedented low of 5 percent in 2008 as nearly vacant office towers sold for record prices.[120]

While critical scholars associate bubbles with the habitual reflexes of capitalists, even mainstream observers recognized that investment horizons had disengaged from the longer-term evolution of the business cycle by about 2005. Over the following three years, even nonperforming loans and vacant buildings were able turn a substantial profit as prices were built on expected (as opposed to actual) occupancy. Lenders used this run-up in values to justify larger loans, expecting the upward trend to continue indefinitely.

Initially obscured by credit-fueled appreciation, the bubble became painfully obvious when debt service obligations started coming due and building incomes were insufficient to pay them. Nationally, foreclosure threatened $400 billion (in 2010) in commercial mortgages, many of them securitized.[121] Securitized loans were some of the first to experience financial distress (compared with loans that were held on banks' balance sheets), as they were larger and their debt service payments consumed more than the underlying assets' cash flow.[122] The performance of CMBS mortgages deteriorated more than that of other loan types.[123] This may be because areas with the highest CMBS delinquencies (Atlanta, Phoenix) were also some of the most overbuilt.[124]

CMBS issuances dropped dramatically when the financial crisis hit in 2008. What had been a $250 billion market in 2007 dropped to 10 percent of that amount the following year.[125] Existing CMBS bonds could not be sold, and new mortgages could not be priced because there was no secondary market for them. Noted one investor, "It's kind of like driving your car after running out of oil, and the engine seizes up. If there's no liquidity and no financing, everything seizes up."[126]

Those closest to the industry suffered first. The liquidity crisis in the CMBS market curtailed new originations, and in 2008 nationwide major office transactions dropped to a quarter of what they were in 2007 (from $207 billion to $50 billion).[127] Between 2006 and 2009 the construction sector lost 22 percent of its employment, almost twice its growth (of 13 percent) during the boom years between 2001 and 2006.[128] After the collapse of Lehman Brothers in 2008, domestic financial institutions and foreign banks with direct exposures to the U.S. market braced for disaster and began shedding employees. The number of mortgage bankers and brokers decreased by approximately 50 percent during the bust (but given the almost 80 percent

increase they experienced during the boom years, they were in a better position than those in the construction industry).[129] Distress in such critical drivers of the U.S. economy stunted overall economic growth; the annual GDP
growth rate shrank in the years that followed. Occupant markets had finally
caught up with the capital markets, but the reunion was not an amicable one.

Financial Times

Developers want to build. In the words of one, "We can't stop. We're always
saying: this is the last project I am doing. But we just can't stop . . . we're like
addicts."[130] The omnipresent machinery, dumpsters, and construction crews
of imported labor are the material enactment of financial capital flowing
through this sector. During periods when capital is abundant, as in the 1980s
and 2000s, developers feed their addictions; they go on buying, selling, and
construction binges that become increasingly divorced from the behaviors of
building occupants.

Given the barriers to liquidity in real estate markets, the extensive and
global integration of the financial and property sectors should be viewed as a
tremendous accomplishment that reconciles fast-moving financial flows with
locally situated built environments. It is the realization of many decades of
political reforms and attempts at institutional rationalization whereby capital
controls and stop valves were weakened, bank activity was deregulated, and
financial innovations flourished. Adding to the mix, the state actively rolled
out its own debt instruments to be bought, sold, and securitized.[131]

During these periods of plenty, however, capital does not switch into all
real estate assets to the same extent. Large financial institutions compete to
underwrite the construction and acquisition of *new* buildings. When they
are flush, tenants and investors trade in their older properties for the latest
models. Their preference for the "new build" is due in part to the everyday
work of nonfinancial professionals who steer capital into particular investments and places. These actors work not for Wall Street or City of London
investment banks, but in backroom offices in Des Moines, suburban office
parks, and City Hall. They are not famous developers and investors like Donald Trump or Sam Zell, but anonymous and modest individuals. It is local
mortgage brokers, real estate brokers, appraisers, building managers, real estate attorneys, planners, and permit expeditors who help create, transact, and
construct desire for the assets that form the basis for complex securitization
schemes. I discuss their roles in depth in the following chapter.

3

Out with the Old: How Professional Practices Construct the Desire for New Construction

Oh, there's plenty of space out there. It's just not where tenants want.
INTERVIEW WITH TENANT A

Financial actors "switch" capital into real estate when returns there are greater than those generated in other sectors of the economy. But why do they also shift capital out of older buildings and commercial districts and into newer ones during such periods? Although renovation occurs during booms, it is overshadowed by the sheer mass, dollar value, and eye-catching novelty of new construction.

In order to understand overbuilding, it is also necessary to understand why actors are so dedicated to what is known in the industry as "the new build." The preference for new buildings is neither intrinsic to developers and tenants nor hardwired into human nature. Instead, it is local real estate professionals who help construct it. These professionals establish and stabilize meanings, making consistent distinctions between buildings. Their distinctions are embodied in market devices, such as classificatory schemata that rank buildings as "Class A, B, and C" and abstractions like "value," which allow consumers to compare buildings and separate the good from the bad.

The work of meaning-making is made to appear technical and naturalistic, not self-serving, by means of two primary techniques. First, professionals regularly defer to the interests of "the market." They treat markets as a remote force of nature, neglecting the fact that they have a hand in determining market behavior and levels. For example, they vigorously promote the notion of consumer sovereignty: the idea that tenants select buildings based on their own pre-formed preferences, over which professionals in the industry have no influence. Second, real estate professionals locate the sources of value (and its antithesis, obsolescence) in what they frame as essential qualities of the commodities themselves. They act as if property enters into circulation with its "meaning defined according to some pre-existing cultural matrix."[1]

Such strategies underplay the extent to which actors on the supply side help stimulate and coproduce demand for the new commodities. Certainly tenants actively judge and evaluate the qualities of different buildings. They test out different options, visit the premises of their competitors, and eventually decide to lease the space they think will best serve their needs. But it is not as if they walk alone in a supermarket stocked with virtually identical products whose attributes can be directly and fully observed. Instead, even the shrewdest tenant is "guided, assisted by material devices which act as points of reference, supports, affordances in which information is distributed."[2] Tenants are swayed by the subtle influence of their advisors and their seemingly objective measures of value. Making a market for new construction, therefore, is joint work involving collaborations between tenants and the suppliers of buildings—not to mention the objects themselves.

Demand is typically latent because moving is expensive and buildings are long-lived. Interactions with developers and real estate intermediaries help coax "replacement demand" out of the tenant base, but do so by appealing to and extending the material interests of the tenants themselves.[3] By ranking buildings, offering financial incentives to move, and denigrating older spaces as "obsolete," these actors lead tenants to believe that they must relocate to new buildings because their former premises will no longer serve their needs.

A systemic bias toward new construction is embedded within the market devices of a critical subset of real estate professionals—brokers, appraisers, city planners, and financiers. These actors make new buildings into calculable and legible assets, which helps overcome some of the obstacles that real estate presents as an investment vehicle (e.g., its social and geographic embeddedness, its opacity, its sluggish production timeline). Designed to absorb global capital surpluses, new buildings become the beneficiaries of recurring bouts of capital switching. In this way intermediaries co-construct demand for new assets that allows the real estate–financial system described in chapter 2 to thrive and reproduce itself.

Cycles do not propel themselves. The practices of real estate professionals are iterative and cumulative: by shepherding the largest and highest-paying tenants toward brand-new structures, they increase the value of such structures, thereby stimulating additional investor demand for more of the same. This progression continues past the obvious break points where there are no more tenants left to procure. If real estate intermediaries were less intent on constructing demand for new products, developers would be forced to absorb the costs of their own excesses and would not overbuild to the same extent.

Brokering

The nature of such needs, whether they arise, for example, from the stomach, or the imagination, makes no difference.

K A R L M A R X , *Capital,* volume 1, 1867[4]

Finance is not an ethereal force that imposes its will on the built environment from above. It comprises myriad on-the-ground economic agents and organizations, ranging from diversified global investment banks to pension funds to trusts to lone day traders. These actors talk to one another, band together in market segments, and keep tabs on one other. But they are limited in their knowledge of specific property markets. They cannot come to consensus about the meaning of abstractions, such as risk and value, involved in the exchange of complex assets and instruments in the absence of local actors. In order for spatially intensive yields to flow from property to investors, an extensive set of ongoing, locally situated relationships must first exist.[5]

For commercial properties, arguably the most important of these relationships is the one between the producers and consumers of place, between landlord-owners and the tenants that occupy their buildings.[6] The price of the most dematerialized mortgage-backed security depends on the income streams of buildings, which are delivered, with some exceptions, by rent-paying tenants located somewhere in space and time.[7] Tenancy is critical because buildings are worth little to developers, investors, and lenders unless they are occupied. When setting values, appraisers scrutinize lease agreements and letters of intent to determine a building's current and future cash flows. Landlords considering the sale of their buildings know that nearly full occupancy will translate into radically higher sale prices. Occupancy, in other words, is a precondition for liquidity at all but the frothiest moments of a speculative bubble.

Ensuring occupancy is the job of real estate brokers. In their capacity as landlord or tenant representatives, they position building products, identify suitable occupants, create marketing strategies, and manage the leasing process. If they represent tenants, they find their clients space and negotiate lease terms for them—getting landlords to throw in valuable concessions such as free rent periods and improvement allowances. In other cases brokers mediate between investors looking to buy and sell their holdings. Without brokers, economic agents would have to transact business directly, incurring considerable search costs to consume what are already idiosyncratic commodities.

In the United States the commercial brokerage profession evolved from the jack-of-all-trades "real estate men" active during the land rushes of the nineteenth century. Real estate boards, made up of honorable members of a new professional class, emerged between 1875 and 1910, and in 1908 formed what later became the National Association of Real Estate Boards.[8] These boards developed codes of ethics, licensing and certification requirements, and standards for common brokerage practices (e.g., lease templates). In the early decades of their existence, they also played a governance role in keeping track of vacancies and exhorting members not to poach other members' tenants.

Despite the best efforts of the founding generation, the technical requirements and certifications for brokering remained relatively lax. Instead, sociability and affect persisted to this day as more important to brokers' success. Since their inception, the culture of commercial brokerage houses has resembled that of college fraternities.[9] Members may no longer have to undergo initiation rites, but they still have to adopt the right attitude, appearance, and jocular demeanor to win acceptance into this status-conscious, mostly male, clique. This is particularly true in the high-stakes world of downtown commercial real estate.

The strong emphasis on sociability emerged because the job of the broker, as a middleman, is to acquire contacts and semi-exclusive information. Despite efforts to make data standardized and more widely available, privileged actors hold monopolies on information such as the fact that tenant X's lease is expiring, tenant Y received two free years of rent to stay in its current location, and tenant Z wants to upgrade a better address.[10]

But brokers were never simply conduits for information; they always helped fashion and standardize user needs, creating desires in their clients through persuasion and salesmanship. Occupants are generally ambivalent about switching their environments. Because their needs are not only material but also aspirational, tenants are highly suggestible. One broker commented, "Most tenants really don't know what they want so we lay out the options for them."[11]

Brokers endear themselves to clients and develop trust through reputation, access to sources of political and financial capital, repeated professional and social interactions, and a similarity of background. Financial firms tend to hire socially similar, or "homophilous," employees, and the real estate field is no different in this regard.[12] Commercial brokers report that in recent years it has become easier to relate to their corporate clients on a personal level because tenants and building management companies are employing a younger, more educated, and more bottom-line-focused cadre of facilities and asset managers in the place of the janitorial staff they used to promote to such ranks.[13]

Drawing on this trust, brokers seek an emotional, visceral response from their clients. Even decisions that could be made according to a cost-benefit calculus seek affective resonance and aesthetic engagement. One commercial broker justified his relocation of a law firm to a new building on the basis not of bottom-line efficiencies, but rather in terms of a more transcendent and intangible kind of well-being: "When they came to their new space with light and air . . . and did a much lighter design, it was like they all needed sunglasses. They just had a pep to their step."[14] I am not trying to draw a moralistic (or vulgarly materialist) distinction here between the "true" and "false" needs of clients, but rather to note that brokers must fuse together meanings (including efficiency and prestige) and collaborate with tenants to successfully position a building.

Brokers convey the more immaterial benefits of place through subtle cues and references to taste, often their own. They drop names—of coveted addresses and architects, of other tenants in a building—that suggest distinction. They employ celebrities or otherwise glamorous individuals to differentiate themselves in a profession where social networks are critical. It is no accident that several of the largest commercial brokerages (The Staubach Companies, now owned by Jones Lang LaSalle; Draper and Kramer) are headed by former college and professional sports legends. The constitutive discourse of the field relies heavily on sports metaphors ("slam-dunks," "pitches"), competitive gamesmanship ("let's *do* this thing" being a commonly heard battle cry), and deal consummation to construct competition in what is a cutthroat, winner-take-all field. Competition not only stimulates aspirational consumption, but also accelerates the already fast-paced tempo of deal making. Brokering, then, like advertising, is a form of commodity production, one "highly dependent on rhetoric, performance, and staging."[15]

The ultimate object of aspiration is the most contemporary building. Corporate tenants "just want to lease space in new buildings . . . New buildings have more cachet, you know, more buzz."[16] Brokers have a material self-interest in promoting new buildings to tenants, as higher commissions and bonuses can be obtained from representing owners of or tenants for gleaming new skyscrapers. In the hierarchy of the brokerage profession, filling new office towers commands the greatest respect and pay. As such, these choice assignments are given to the most accomplished and reputable agents.

But getting tenants to reevaluate the qualities of, and eventually "detach" from, their current space is no small feat.[17] Brokers try to appeal to bottom-line concerns, emphasizing that modern building materials and designs will allow tenants to save on utility costs, or that concessions will bring direct rental costs more in line with those of older buildings. They also encourage moves by making affective appeals to their status-sensitive clients: "Like developers, brokers are seductive, charismatic, and attractive. They envelope the tenant in a grandiose vision: 'Imagine that you can put your name on that new building!' Their job is basically *to create a feeling of need* [italics mine]."[18] One broker noted, "Your job is to make them [tenants] feel dissatisfied with where they're at now. You convince them they have a problem and then, *voila*, you solve it for them."[19]

This fueling of aspirations is helped along by the neutral-seeming ranking system used to standardize income-generating assets. Starting as early as the mid-1920s, brokers and building managers segmented commercial buildings into three classes: A, B, and C.[20] Office buildings, for instance, are associated with a distinct set of characteristics:[21]

- Class A: Landmark quality, high-rise buildings with prime central business district locations; generally 100,000 square feet or larger (five or more floors), concrete and steel construction, built in the last ten years, business support amenities, strong identifiable location and access.
- Class B: Older, renovated buildings in good locations.
- Class C: Older, unrenovated buildings in average to fair condition.

The rankings are intended to synthetically represent the qualities of buildings and therefore the rents that landlords can charge for them. They therefore reflect the degree of status and commodification possible. One investment broker noted, "When we talk about 'tradeable inventories,' we're really just talking about Class A buildings, occasionally some Bs."[22] The names of "starchitects" such as Calatrava, Hadid, and Gehry arose in conjunction with the argument that "only global buildings appeal to global tenants and investors."[23] Despite their subjectivity, these typologies help tenants make relocation decisions, and investors rely on them as signs of a building's quality and ability to generate cash flow.[24] Boiling down something as complex as physical and locational quality into a simple three-part rating system is a way to create cognitive tractability from afar.[25]

In contrast, Class B and C buildings are not global buildings, their markets being almost entirely local or regional. Buildings that have endured through the decades are considered idiosyncratic insofar as local knowledge is required to understand why the floor plans are quirky, the office spaces small, and the parking on-street. The older a building gets, the more difficult its resale value is to predict. Because these buildings are considered difficult to unload in the global investment marketplace, they are lower in the pecking order used by brokers. Only a select handful of trophy buildings from previous booms convey a sense of lasting value, even those that have been significantly and visibly retrofitted.

While their affective appeals may give the appearance that brokers tailor their advice to the individualized needs of each client, they nevertheless promote relatively generic distinctions between buildings. Specifically, they help manufacture consensus about the superiority of a small number of buildings; for global tenants, those buildings are the new, Class A towers in a select number of cities and suburban office markets. In this way brokers validate new buildings by co-constructing an occupant and investor market for them—even in the absence of growth and expansion. One broker wrote off concerns over adding to a saturated market: "These new Class As are so good, you'll see, tenants will come out of the woodwork for them."[26] In contrast to the textbook lesson of demand determining supply, brokers help suppliers of new space construct their own demand.

Appraising

Happily there is nothing in the laws of value which remains for the present or any future writer to clear up; the theory of the subject is compleat [sic].
JOHN STUART MILL, *Principles of Political Economy*, 1848[27]

Challenging Mill, I would argue that the notion of "value" is still far from straightforward. The valuation of property is required for all private debt and sales transactions as well as for public condemnations and property tax assessments. The single value appraisers produce is intended to reflect what an uncoerced purchaser is willing to pay for a property as well as the risks it poses to investors and lenders (i.e., the price at which they would be able to sell the asset in the event of default). But market value is revealed only when the property is actually purchased. Moreover, it is relative, which is why appraising involves so many acts of comparison—the establishment of equivalencies and distinctions between structures and neighborhoods. Is a house with two bathrooms and a European-style kitchen equal to a house with three bathrooms and a retro kitchen? Appraisers abstract and convert radically different qualities into a singular numerical representation that allows assets to be compared.

Numbers are intended to convey objective market information, free from the distortions and ambiguities of social context.[28] Still, the appraisal profession has struggled to assert its authority to divine the "true" value of property.[29] In response to questions about its neutrality and professionalism, appraisers created quasi-scientific techniques, drawing on three methods of observation and analysis developed in the early decades of the twentieth century.[30] The first, the "cost" approach, derives an estimate of value based on the replacement or reproduction price of the real property, minus depreciation. The replacement price reflects the land, labor, and capital necessary for the construction of a property considered similar to the one in question. The second, the "comparable sales" approach, derives a value from transactional evidence for similar properties. The third technique, the "income" approach, is based on conventional discounted cash flow analysis and the use of a capitalization (or "cap") rate that converts income into an estimate of value.

These three techniques can produce three separate estimates of value for the same property, which appraisers must then "reconcile." The alchemy of reconciliation requires appraisers to use their experience and sensitivity to local context to inform their opinion. As such, value is exposed to the subjective affinities of appraisers and their clients for particular places and aesthetics as well as discrimination against those buildings and places that produce discomfort.[31] Particularly in the absence of active markets and comparable building types, appraisers fall back on their social and professional contexts— following the herd, relying on stereotypes, or allowing their clients influence.

Although appraisers disparage brokers as "inferior analysts but superior networkers," such attitudes misrepresent the degree to which appraisers, like their broker cousins, rely on networks, status, and social cues in reconciling their valuations.[32] For example, valuations for major portfolio acquisitions are conducted by the largest and most reputable firms. And like brokers, appraisers convey their trustworthiness and expertise through signals—listing the alma maters and majors of their employees (inevitably, economics) and the names of the "silk-stocking" banks and investment firms for which they do business. Observed one, "They [our clients] are image-conscious. And so are we."[33]

Such deference to image and power reveals itself in appraisals for new commercial buildings. New construction habitually receives higher appraisals than older buildings, for many good and not-so-good reasons. Building age is perhaps the most common input, other than size, in all three appraisal techniques. Although dates are easy to establish, "old" and "new" are subjective qualifiers; most buildings are hybrids, having been "modernized" at some point after the original construction. If a renovation is not highly visible, it is unlikely to have been recorded and widely known. Noted one appraiser, "We generally try to compare new buildings to new buildings and old buildings to old buildings, but it's sometimes hard to tell."[34]

The appraisal of more contemporary investment properties typically relies on the income approach, in which a stabilized year's net operating income is divided by a market capitalization rate, or on the discounted cash flow approach, in which multiple years of net operating income are converted to a single present value. Both methods depend on an accurate rendering of a building's income, which is difficult to know if tenants are receiving reductions in their rents as part of concession agreements. Lease documents may not account for concessions (particularly if they have already been disbursed, i.e., rent abatement periods passed and improvement allowances paid), as such information is heavily guarded. When concessions are provided to attract tenants to new buildings, rents there will appear higher than they are in reality, and the buildings' values will be overstated.

Moreover, the cap rates that appraisers use in the income approach are generated by acquisition markets that are driven as much by capital gluts and stock prices as they are by leasing income. The vast quantities of debt and equity available to purchase new buildings during bubbles have bearing on appraised values. To an appraiser, a building sold at the peak of a speculative bubble is a still considered a comparable, even if the purchaser covered the full purchase price with cheap CMBS debt. In contagion-like fashion, inflated estimates quickly become "reasonable" data points for future appraisals, pushing up the values of similar buildings. In this way appraisals sway the very market they are assumed to passively reflect.

Appraisers work for banks and investment firms, and they mimic their financial clients' underwriting standards and preferences for the new.[35] Noted one appraiser, "[The banks] will set out a 'range of reasonableness' for us. But the final value has to be supportable by our numbers."[36] In other words, clients tell appraisers where they would like target values to ultimately settle, and appraisers find evidence to justify those values. Because clients often lack the time to visit properties and ignore problematic or ambiguous information, appraisers provide them with final values that reflect their preset opinions. While appraisers defend such imitation as a sharing of information between client and service provider, other practitioners joke that MAI, the acronym for the professional distinction accorded by the Appraisal Institute, stands for "Made as Instructed."

In addition to establishing values, appraisers promulgate standards for identifying the antithesis of value, obsolescence. Although the term had been in common usage since the early 1800s, it was only during the early 1900s that it gained a technical legitimacy in the United States.[37] Appraisers, lawyers, and economists in the burgeoning real estate field devoted chapters of their new textbooks on the principles of real estate and land economics to the topic.

The interest in obsolescence in the early twentieth century reflects a high level of anxiety about the fact that buildings of the recent past—many of which just twenty years earlier boasted the latest designs and technologies—were being plundered of tenants, abandoned, and in many cases demolished. In some cases marginally older buildings were torn down so that taller, more profitable structures could replace them. But even in growing cities such as Chicago, New York, and Philadelphia, some were razed for parking lots, low-rise garages, and low-density commercial buildings called "taxpayers" because they paid the property taxes and that was all. Before this period, it was less common to demolish a building because it was outmoded; owners just built on modern additions.[38] As such, architects and engineers puzzled over the impermanence of such complex human endeavors.[39] The general public was also confused: while the "wear and tear" and physical impairment of a building caused by use was visible to the untrained eye, the argument that buildings needed to be demolished because they had fallen out of style or were supplanted by more modern alternatives invited skepticism.[40]

Appraisers argued that marginally older buildings could still experience "financial decay" despite their physical health and asserted that obsolescence was real and calculable.[41] It was "a falling off in the value or usefulness of a thing from causes outside of the thing itself, as opposed to the effects of wear or physical deterioration."[42] As such, obsolescence became an economic category reflecting supply and demand conditions and not structural (engineering and architectural) ones.[43] Buildings could become obsolete not because they were unusable, but because they could not be used as profitably as current or future investors desired.[44] Obsolescence is reflected in a "loss in property value above normal age or physical depreciation," and appraisers quantify it by comparing a property to its highest-performing comparable.[45]

Appraisers developed understandings of a building's usefulness in terms of standard life "spans" and "expectancies." The scholarship from the 1920s and 1930s is almost Gothic in its obsession with death—the inevitable end point of a building's life. Grandfather of the real estate profession Frederick Babcock argued that building improvements were ticking time bombs whose values began declining immediately after construction.[46] But the exact moment of implosion was hard to predict. The president of New York City's Tax Department observed that "some buildings in Manhattan become obsolete in five years and cumber the ground."[47] The majority of scholars pronounced that office buildings had a "useful and profitable life" between twenty-five years and forty years. Once they hit this mark, such buildings would become candidates for demolition, modernization, or conversion to other uses. Some could persist past their time, typically those "enshrouded with a halo of tradition and sentiment which offsets the effects of obsolescence."[48] But otherwise obsolescence was treated as a condition caused by the naturally cyclical laws of development and as unstoppable as the winds of time.

Appraisers made obsolescence into a boon for property owners, arguing that no one should have to pay taxes for the loss in the economic value of their holdings caused by obsolescence. They fought to have a depreciation deduction included in the income tax code so that owners could receive an annual offset in gross income equal to the portion of the asset's depreciable life that had transpired and the amount of "normal" obsolescence that had occurred.[49] Calculated on the basis of a building's appraised value, the deduction continues to be most advantageous to owners of new, highly valued buildings—an irony as aging benefits the young more than the old.

Improvements are not the only commodity susceptible to obsolescence. The other factor in the value of real estate—land—also loses its luster over time as the locations of business districts shift. In Chicago, for example, the main commercial artery in the Loop moved from its pre-fire location on the east-west transverse of Lake and Washington Streets to the north-south of Dearborn Street in the 1880s and 1890s. It then moved to LaSalle Street in the 1910s and 1920s (and eventually out to Wacker Drive in the 1980s and 2000s booms). Each mass relocation impaired values in the areas left behind; many believed that any subsequent investment there would be "wasted."[50]

Appraisers in the twenty-first century continue to frame obsolescence as the inescapable by-product of autonomous locational and technological change, progress that should be lauded and not lamented. Noted one, "Better performing buildings will naturally make the older ones obsolete. That's the beauty of our system. Older buildings are pushed down the value chain and into other lesser building types and markets where they'll be of interest to tenants with smaller budgets."[51] Such progressive change is inevitable and "as certain and ever present as the forces of nature."[52] Whether the building you own falls inside or outside the boundaries of the market is just the luck of the draw.[53] This naturalization removes the agents of change, obscuring the political forces and social prejudices that build up certain places and devalue others.

Real estate intermediaries also continue to use the construct of obsolescence circuitously. In some instances, obsolescence is said to result from the natural urge of developers and tenants to innovate. In others, obsolescence is seen as an innate attribute of older products, as if the impaired valuations of aged stock are proof of a disease that plagues buildings and submarkets *as a result of* their age.

The idea that buildings are essentially dead before they are born satisfies the need of the real estate industry for a never-ending construction pipeline. If buildings are always aging and depreciating, developers can justify their decisions to build even when market conditions are not favorable. They can discount overproduction if the scourge of obsolescence is constantly "attacking" existing buildings.

City Planning

Since the mid-1960s the increasingly dominant and exclusive strategy of the
capitalist state is to solve the problem of the obsolescence of the commodity form
by politically creating conditions under which legal and economic subjects can
function as commodities.
CLAUS OFFE, *Contradictions of the Welfare State*, 1984[54]

New opportunities for value extraction may arise from the ashes of the devalued as developers buy low, (re)develop, and sell high. But if the private sector is slow to realize these opportunities or if demand is sluggish, local governments may step in. The capitalist state recalibrates the supply-demand balance, simultaneously encouraging supply while tightening up those markets coping with a surplus of older buildings. Local governments lend their legal authority to demolish or stimulate the conversion of excess capacity by labeling it "blighted."

At the same time the nascent field of real estate was developing the techniques that would come to define the industry, its professional associations and real estate boards were urging members to involve themselves in public planning issues:

> As an association you should concern yourselves with anything that affects
> the growth and development of your city, more especially its downtown section. City planning, zoning ordinances, railroad terminals, street car facilities,
> boulevards or anything else that tends to limit, control, modify or develop the
> business district is of more vital significance to the investment in your property than all of today's operating or renting problems put together.[55]

They also called for more, not less, government intervention into property markets, conceding that the state should lay the infrastructure and legal framework for developing new areas and redeveloping older ones.[56] As the field of city planning began to take shape midcentury, vocal elements of the commercial real estate sector called on it to become more aggressive, particularly in terms of downtown commercial development. Initially such requests fell on deaf ears: planners assumed that a thriving central business district was an immutable feature of cities.[57] But as early as the 1940s, corporations were moving away from their historical downtown locations. Arguing that cities needed to compete with both their own fast-growing suburbs and rival regions, private actors persuaded municipal government planners to absorb some of the front-end costs of new development and to eliminate the older structures that stood in their way.

Local governments also pursued their own interests—in status, growth, and fiscal stability—through commercial redevelopment.[58] A growing tax base allows for ambitious public improvement schemes, financed by the revenue extracted from it and from the debt made cheaper by higher bond ratings. The accomplishments and dictatorial excesses of individual city builders such as Robert Moses (in New York City) and Ed Logue (in Boston) are well known in this regard. Elected officials often measure their worth by the number of cranes in the sky and are eager to treat any building boom within their jurisdictions as validation of their pro-growth policies and plans. Their short-term horizons lead them to take what they can get when they can get it; concerns about empty buildings or the solvency of overleveraged developers can always wait until the next administration.

Below mayors and councils of elected leaders sit the authorities, planners, and consultants that accomplish similar objectives more incrementally but typically fly below the radar. Municipal redevelopment authorities, for example, have a focused real estate agenda and access to finance beyond traditional revenue sources.[59] They buy and sell property, acquire property through eminent domain, issue redevelopment bonds that are not subject to popular votes, and grant tax relief to encourage redevelopment. Some cities have opted to keep these same powers in-house, empowering planners in agencies for community development and housing and adding the personnel necessary to execute increasingly complex property and debt transactions. Still other cities prefer to bring in specialized knowledge by contracting out to planning consultants.[60] Together with other major city institutions such as universities and the media, the public sector constitutes the urban "growth machine" that dominates development politics and governs the amount and nature of new supply.[61]

A common feature of the activities of the growth machine in the postwar years has been the underwriting of creative destruction. It remakes space by championing new assets while eliminating older "obsolete" ones.

On the creative side, local governments are accomplices in the periodic buildups that transform cities. Figure 3.1 reveals the interlocking fortunes of local governments and real estate professionals as public sector revenues and employment rise and fall with commercial real estate markets. During booms, increased property, sales, and income tax revenues fill the public cof-

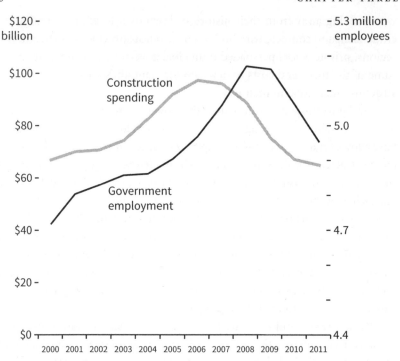

FIGURE 3.1. Local government employment and total construction spending, United States, 2000–2010. Data from U.S. Bureau of the Census, Annual Survey of Public Employment and Payroll; Federal Reserve Economic Data (FRED), Federal Reserve of St. Louis. Construction spending is not seasonally adjusted.

fers, enabling the implementation of previously conceived plans and ad hoc deals. New construction is more fiscally productive because such buildings capitalize their ground rent with higher densities and tend to be assessed for taxes at higher values per square foot. Moreover, developers are some of the largest contributors to the reelection campaigns of officials who grant zoning and subsidy requests.[62]

City governments rely on three primary mechanisms for accelerating new construction. First, they can relax the regulatory burden placed on developers, owners, and investors. Planners can loosen construction requirements (permitting and entitlements), provide tax relief, and upzone to allow for more intensive use of targeted areas. Second, they can use the power of public investment to enhance the value of new construction or underwrite its financing.[63] Local governments can site new transit stations, upgrade infrastructure (such as widening streets), and use eminent domain powers to condemn and convey private property to encourage investment in new commercial districts. They also offer low-interest loans and equity financing for new construction. Third, cities can promote "hot" new areas and buildings, using imagery and branding to encourage consumption.[64] The subsidies, amenities, and other inducements supporting construction represent a kind of political consensus about new buildings and provide both material incentives and intangible signals for developers to build.

Less recognized but no less important is the role played by local governments in the destruction accompanying such creativity. Cities are responsible for clearing out the excess capacity left behind by previous waves of accumulation.[65] They use public funds to demolish older structures and convert older uses, particularly in areas where land values are growing.[66] Or they choose not to defend these structures, allowing private developers free rein to raze them. Since the 1980s cities have focused mainly on eliminating the ruins of the Fordist industrial economy: the idle manufacturing plants, empty warehouse facilities, and infrastructure (e.g., freight lines) that supported and connected them.[67] Surplus property classes vary by city, and they may also include Class C office buildings whose tenants have relocated and institutional buildings whose user populations have shrunk. If they cannot be eliminated, then obsolete properties are, at a minimum, contained. Through public investment, regulatory policy, and the creation of distinct submarkets, the state enforces boundaries that "restrict the effects of devalorization, economic decline, and asset loss to clearly circumscribed neighborhoods and protect the integrity of mortgages in other areas."[68]

Confining or removing obsolete properties neutralizes the harsh effects of overproduction on individual owners. In his formulation of "class monopoly rent," Harvey points out that property owners benefit by acting collectively to create artificial scarcity (e.g., by running housing units down and taking them off the market).[69] Tighter markets register greater appreciation; without the slack provided by the presence of less desirable assets, available demand can more quickly soak up any new supply.

Seen in this light, municipal planning is as much about the removal of excess capacity in the face of structural oversupply as it is about creating jobs or public amenities. Investors would rather not wait until a factory falls to the ground or its workers retire to realize a development opportunity. They would prefer for the state to jump-start this process with forced devaluation, condemnation, or destruction of older assets so that such impediments do not block their path. Removing such roadblocks facilitates "internal" capital switching: investment can move between old and new products more efficiently when the costs of redevelopment are shifted to the public sector.

Like brokers and appraisers, local governments discursively code the meaning of place through policies and practices that can be highly advantageous to capital.[70] Specifically, they use their legal authority and their ability to shape public opinion to create consensus about the location of value and obsolescence. A long-standing use of state power in this regard has been the designation of places as "blighted."[71] Codified in the urban renewal statutes starting in the 1940s, a "finding" of blight had to be present in a particular submarket to justify the use of police and expenditure powers to benefit private property owners there.[72] Blighted areas were defined as slums, or "areas in which land values after a period of increase are stationary or falling, areas in which buildings have become more or less obsolete, or areas characterized by building vacancies and the appearance of decay and dejection, where there is no prospect of a renewed market for its original use or for other purposes."[73]

Initially blight was seen primarily as a threat to the use values of property; it produced hardships for residents and bred crime and disease.[74] The fact that blight was disproportionately "found" in nonwhite areas did not rattle early planners.[75] Business interests and their allies in City Hall soon realized that the government's finding of blight could aid in their intention to redevelop those commercial areas surrounding downtowns that included few residents, and the meaning of the term began to merge with that of obsolescence, which is a threat to exchange values. The terms "blight" and "obsolescence" are complementary: obsolescence is mainly used to refer to an individual building, and its symptoms are mainly economic, whereas blight is perceived as an attribute of an entire area and has social associations built into it.[76] While obsolescence is primarily a private affair, blight requires state action. A blighted area was originally defined as "a district which is not what it should be."[77] As with obsolescence, the definitions remained vague and tautological: blight was a cause of physical deterioration, a state of being in which the built environment was deteriorated or physically impaired beyond normal use, and a future possibility. This confusion and the imperative for city growth in the face of suburban competition kept the courts from using a standardized definition that might have prevented some cities from indiscriminately assisting profitable redevelopment.

Planners turned to technique to fend off criticism and lawsuits. Starting in the 1930s and 1940s, state urban renewal statutes, and subsequently local ordinances, attempted to pin down a more exact meaning of blight by listing "blighting factors" that had to be found in a neighborhood before public funds could be committed.[78] These factors included the age of buildings (anything over the "useful life" of thirty years was in trouble), density, population gain or loss, lack of ventilation and light, and structural deterioration. Planners drafted maps that depicted, to the block level, the epicenters of decline and areas of potential blight. For example, in 1943, the Chicago

Plan Commission designated a twenty-three-square-mile corridor around the central business district as blighted and in need of public assistance.[79]

During the urban renewal era, clearance was the favored remedy for blight as planners purported to sanitize space and eliminate the possibility of the blight returning by removing the infected buildings completely.[80] With suburban "greenfield" sites setting the standard, vacant parcels were thought to be an easier sell to private developers than parcels with devalued structures left intact.[81] The federal government reimbursed cities for two-thirds of the expense to condemn, purchase, demolish, and resell surplus private property. Federal funds subsidized developers and unionized construction workers with cleared prime land at bargain-basement prices (i.e., "write-downs") and displaced over a million people.[82] In order to support the bulldozer approach, cities actively campaigned against rehabilitating older properties, labeling such activities wrong-headed and illegitimate.

To justify the continuation of revenue transfers to developers after urban renewal funding ended in 1974, municipalities adhered to the well-worn strategy of identifying obsolete properties with potential to be removed or redeemed through public intervention. For example, states kept the blight requirement in their enabling legislation for tax increment financing (TIF), the most commonly used local economic development tool in the country in the late twentieth century.[83] TIF enables municipal planners to designate districts for redevelopment and to use the expected increase in property taxes from the redevelopment (the "increment") in those districts to pay for initial and ongoing expenditures related to the redevelopment of property there.[84] The professional opinions of appraisers and consulting firms establish that the blight requirement for each TIF district is met by tabulating those features of the area that meet the legislation's standards—most of which have not changed since they were included in postwar urban renewal statutes. These features include the number of buildings over the age of thirty years, obsolescence (typically left undefined), and depreciation or lack of physical maintenance.

TIF funds are still regularly used for demolition, land assembly, and below-market sales of land to private developers for new construction. However, clearance—that is, the outright demolition of older structures—has not occurred on the same scale as it did during the immediate postwar period. Local governments have found less destructive ways of removing excess capacity, including adapting obsolete buildings for alternative uses. For example, the City of Los Angeles considers buildings five years or older as candidates for publicly subsidized reuse, and under the Lower Manhattan Revitalization Plan, New York City offers zoning accommodations and subsidies for converting older office buildings into apartments.[85] Conversion strategies for surplus property have grown in popularity as municipalities designate more of their stock as landmarks and their economies shift in orientation from production to consumption.

The blight justification for public intervention in property markets allows local governments to subtly shape the gray area that separates public purposes from private ones. When blight is conflated with obsolescence, structurally sound buildings in areas hard-pressed to be called deleterious to one's health, safety, and morals can be redeveloped with public money. After all, the richest development opportunities are often not in truly marginal locations, but rather in or near places where capital is already circulating ("the blight that's right"), such as the resurgent downtown cores of cities like Chicago.[86] This conflation also justifies pro-cyclical public interventions, as when cities subsidize developers during periods awash with private credit.

By lagging behind developers, however, cities often enter the redevelopment game too late. When they do, they help extend booms past their limits, converting them into bubbles with the risk of overbuilding. Subsidies for new construction and public improvement schemes get capitalized into the appraised values of private property, buoying appreciation in areas where speculation is already under way.[87] Eliminating the surpluses left in the wake of booms similarly encourages developers to keep the new construction assembly line moving.

Financing

> I mean, look at the Empire State Building: its floor plate is so narrow. Those older
> buildings are gorgeous, but they're just not designed to cash flow.
> INTERVIEW WITH LENDER B

With few geographic barriers to capital mobility, investors around the world have free rein to buy those assets that best meet their spatial selection criteria and to ignore those that do not. Each segment of the financial sector has its own risk-return profiles and favored cities, neighborhoods within cities, and building or product types. These preferences, like those of tenants, are derived from their own evaluative criteria as well as from their interaction with the local professionals described above who act as matchmakers and advisors in specific property markets.

Loan officers and investment portfolio managers do not tell developers when, where, or what to build. Instead, their underwriting standards, far more than mere technical argot, reflect institutionalized predispositions for particular kinds of assets that financiers assume will contain risks, appeal to the most conventional and deep-pocketed tenants, and make sense to other investors and securities holders further up the food chain.[88] Before they even submit applications for project financing, developers internalize the evaluative criteria of funders. For example, developers seek to maximize the amount of rentable building space so that each square foot is financially productive enough to justify the cost of acquisition and construction.[89] As Lender B noted:

> We don't make them [developers] go back to the drawing board if we're not happy with their designs. They just know enough to give us what we want . . . There are certain kinds of projects that we would never fund, and they know what those are. Developers cater to us by designing projects to be attractive to us on the revenue side. It's not such a stretch . . . in most cases, it helps them too.

In other words, the "ask" is rigged because the professional standards for a cash-flow-maximizing building are established through repeated interactions between the property owner and sources of capital. By selectively funding the development and acquisition of certain kinds of projects, adopting particular discount rates and appraisal methods, and requiring that proposals meet specific underwriting standards, financiers signal to the development community what is and is not a financially feasible project. Supply is tailored to the risk and return requirements of capital; sometimes these are aligned with tenant needs, and at other times, they diverge.

Large investors generally stick to standardized locations and products as they look for buildings that they can sell easily. They seek two criteria at a minimum: "global gateways" and "institutional-grade" properties. These are code words for places and assets believed to hold their value and appeal to risk-averse institutional investors. Due to the volume of capital they control, institutional investors tend to form the baseline, moving pricing and guiding the industry toward particular assets.[90] Developers seek to satisfy the requirements of their direct sources of capital and indirectly of the aggregated consumer and investment "markets."

In North America the institutional segment of the capital markets considers the largest and most expensive cities the "top tier" for commercial real estate: New York City, Washington, DC, Boston, San Francisco, Los Angeles, Toronto, Vancouver, and Chicago. These are cities where pricing has been among the highest in the world and where returns are considered "reliable."[91] For the most part, these cities have modest zoning and building restrictions, with enough choice building sites to allow for new construction but not so many as to dampen any appreciative effect from it, so that investors are able to capture quasi-monopoly returns.[92] At the same time, global investors target areas where developers can erect buildings quickly, such as sprawling suburbs and Sunbelt cities with low regulatory hurdles, such as Dallas, Atlanta, Phoenix, and Houston.

Within these urbanized regions, institutional capital is drawn to those tested microgeographies where it can capitalize on the benefits of agglomeration. Particular addresses command premium rents and land values: for retail space, Rodeo Drive in Los Angeles, Fifth Avenue in New York City, or North Michigan Avenue in Chicago; for office space, Wall Street in New York City, L Street in Washington, DC, or Wacker Drive in Chicago. This list is in constant flux, however, as new areas warm up while others fall out of favor—thanks, in part, to the work of smaller, risk-taking investors and the positioning work of brokers and planners.

Local knowledge of pricing and leasing dynamics within such submarkets is more valuable to distant investors than aggregated data. Noted Pension Fund Advisor A, "We might buy in a metro that looks oversupplied but we're buying in a submarket that's always tighter than the metro." Only brokers based in or deeply familiar with the locale possess intelligence of this kind. During booms and bubbles investors are willing to pursue more aggressive submarkets and sites that inch toward the edge of fully built-out areas and break new ground. However, they typically retreat to the core when capital is scarce and prices fall.

Investment in real estate must deliver profits to its legal beneficiaries far removed from any particular site in what is essentially a widespread form of absentee ownership.[93] The local knowledge needed to understand less tested submarkets is expensive to procure and often fallible. The assets themselves, therefore, must be easy to fathom not just for occupants and lenders, but also for the purchasers of securities built on debt secured by the property. Ever since institutional investment and securitization in real estate became commonplace in the 1980s, assets have been expected to perform over progressively shorter time periods. As the chain of ownership becomes more attenuated, fragmented, and subject to quick trades, the onus is placed on the assets themselves to be quickly calculable and legible to distant investors.

In order to travel, assets are stripped of their context and abstracted from local conditions. Their physical form faces pressure to become equally abstract.[94] As a part of this move to standardize, global investors tend to prefer generic new construction over older buildings with specific, often complicated histories and quirkier physical features. The portfolios of institutional investors are weighted toward new buildings.

New buildings can be traded more quickly by those lacking local knowledge and spatial literacy. Uninformed investors gravitate toward uniform signifiers of value—namely, toward the newest, largest, best located, and most marketable and amenities-rich space. For the seemingly banal reason of reducing the risk of nonpayment, they fall back on formulaic product types and have limited tolerance for anomalous development. Real estate writer Chris Leinberger complains, for example, that investment banks consider a total of only nineteen narrowly defined real estate product types.[95] A "neighborhood retail center," for instance,

> occupies 12 to 15 acres, anchored by a supermarket/drug store of between 50,000 and 70,000 square feet. It also includes in-line stores of national chains and franchises. The buildings occupy 20 percent of the site and are set back from the street; the balance of the land is surface parking. The location has a minimum of 20,000 people living within a three-mile radius and will have demographic characteristics appropriate for the particular supermarket chain. The center will be sited on a street with at least 20,000 cars per day passing by. It will preferably be on the "going-home" side of the street.[96]

Any deviation from the norm is likely to jeopardize not only the development's appeal to national retailers but also the project financing.

One of the easiest ways to gauge a building's value at a distance is to scrutinize its tenant base. New structures are most likely to house major national and international corporations—in other words, "creditworthy" tenants. Reputation and sales matter, but a tenant's access to financing is considered one of the best hedges against future uncertainty. Noted one broker, "If you're an owner, you'd rather have GE [General Electric] as a tenant than Google. I know it's unexciting but GE has stable cash flows and the credit of the tenant means a lot . . . especially when it comes to a reversion [sale]."[97] The financial status of tenants, their ability to access debt to fund operations, is capitalized into the value of the building. The "preference" of tenants for new buildings, therefore, is also a function of the amount of credit flowing through the system.

Investors argue that their predilection for contemporary buildings is based on the fact that their physical forms are designed to financial standards in ways that those of older buildings were not.[98] This is a common and historically grounded trope.[99] Skyscrapers in New York City and Chicago built during the booms of the 1880s and 1920s were also influenced by the interplay of financial pressure and local regulatory conventions. The need to maximize cash flows produced nearly identical floor plans in both cities despite distinctive exteriors (Chicago with shorter, blockier towers and New York City with tall and lean buildings erected on smaller sites). The 1920s witnessed "the rise of a new kind of building, designed and built for the express purpose of maximizing rents from a varied multitude of tenants and turning a profit for the developer."[100] In recent years, financial pressures have led developers to try to minimize the amount of space "wasted" on non-revenue-producing uses such as common space. The standards for what is considered a financially optimal building change over time as they are tracked to the willingness of occupants to pay for specific features. Every generation of developers thinks its buildings are the most efficient and finds something to criticize in the buildings of previous eras.[101]

Even though surveys have revealed that building age is less important to tenants than the configuration of space, developers and investors argue that older commercial buildings cannot generate sufficient cash flows, possess too much interior and idle space, and were not designed to exploit the heights they often achieve.[102] Despite their beauty, their massing is "incredibly inefficient from a financial perspective."[103] When asked about the possibility of lending money to modernize 1920s-era loft office space in Chicago, one lender replied,

> God, we wouldn't touch a retrofit like that with a ten-foot pole . . . With older buildings, you don't know what you're going to find once you start construction. Also the tenant mix is less desirable. Clients in that kind of buildings are going to be smaller, lower paying . . . they don't have the balance sheets of large corporations. We wouldn't even go near a landmarked building . . . New construction is just much more straightforward.[104]

Modernization is not an option for many older buildings beyond fixing immediate threats to their income streams. Instead, financiers prefer to channel their profits into new construction. Although renovation typically costs half as much as new construction, getting in early on a new building can generate higher returns.[105] Noted one developer, "Investors can save money by buying existing assets, but if they want to get 12 to 15 percent returns, they've really got to get into the new stuff."[106]

The underlying financial rationale for such (dis)investment decisions rests on the fact that rents cannot be raised as easily in older buildings where established, credit-poor tenants have long-term leases. Only a highly visible "gut" rehabilitation or conversion to a more popular use would convince existing tenants to execute higher leases or attract a new cohort of tenants, but such strategies are risky (the larger tax bill, on the other hand, is more certain). Moreover, investors believe that operating expenses in older buildings are higher than for comparable newer buildings.[107] Even if expenses vary, current costs are more quantifiable and assured than projected, future rents. The short-term horizons of most financial professionals also lead them away from buildings that cannot guarantee capital gains from an imminent sale.[108]

Older buildings, therefore, have a harder time meeting the financial specifications of investors, do not trade as easily, and cannot be converted as readily into financial paper. A loan officer at a foreign-owned bank scowled at the prospect of lending for a major rehabilitation: "We would only do these deals as a favor to someone. And it would have to be a big favor."[109] Banks' underwriting criteria embody a bias against older buildings, which are perceived to be greater credit risks. There is some basis for this perception, as empirical studies correlate increasing building age with lower investment performance and sale price.[110] Whether such findings validate the idea that older construction is inherently less desirable—or whether financial actors are just less likely to provide capital to maintain their value—is an open question.

Owners of older buildings have no choice but to seek out more patient capital from small equity providers and regional banks. Restricted access to well-capitalized financiers condemns older buildings to low profit margins and rates of reinvestment, high operating expenses, high-cost debt, and the prospect of diminished utility. If more local knowledge is required, either profits will have to be higher in order to compensate for these inflexibilities or other parties, such as the local government, will have to assume a portion of these additional costs.[111] In what is essentially a two-tiered system, high finance and global development partners chase new trophy projects, leaving older commercial buildings, landmarked structures, and reuse projects for smaller developers, regional banks, nonprofits, and the public sector.

Too Much of a New Thing

Along with tenants and other members of the real estate industry, the professions discussed above influence the demand for specific products, making a market for new construction—partly by denigrating the utility of the inventory that pre-dates it. Their ability to weigh in credibly on questions of value and their use of standardized typologies allow diverse properties to be compared with each other, ranked and prioritized, and finally, priced and consumed by occupants and investors. The meanings they affix to property are enduring and self-fulfilling. For example, the single number issued in an appraisal report or the mark of a Class A building are claims that are inscribed on the asset as it passes through subsequently more complicated stages of its financial life cycle. Similarly, when a building or neighborhood is labeled as

obsolete or blighted, the stigma sticks until public assistance, private investment, and visible "reinvention" wash it off.[112]

These practices create the conditions for the financialization of real estate.
Real estate professionals, therefore, are not simply advice givers or knowledge brokers, but also key actors in the production of financial commodities.
Their techniques for classifying property are neither representations of "real"
qualities nor smoke screens for capital accumulation, but "actively operate to
make possible the stakes, dispositions and values of the field itself."[113] Some
scholars would go so far as to say that they "perform, shape and format" the
markets they ostensibly describe.[114]

Although they work in different sectors and through different media
(affect, regulation, technical expertise), each of these professional groups
reinforces the others' work though repeated transactions and through dependence on one another for the successful execution of their contractual
commitments.[115] For example, property values appreciate as a result of the
recursive feedback of different professionals across the industry acting on
comparable hunches at roughly the same time. Norm followers are rewarded
for this kind of "mimetic rationality" with higher commissions, tax revenues,
and investment returns.[116] Incentives to collude reinforce mutually consistent narratives of when markets are cresting and waning. The coordinated
calculative behaviors of a few key professionals and institutions will attract
other members of the industry to and away from similar properties and submarkets at roughly the same time, "performing" cycles.

The herding impulse is also a response to the speedup of the entire industry created by boom-time demand for financial commodities and accelerated
production schedules. Bringing a new bond issuance to market or executing
a purchase creates an urgency that compels other professionals to follow the
crowd. During such times the collective judgments of others are believed to
be better informed than the individual ones professionals could make on the
fly. Brooking a trend and forging out on one's own takes more time and effort.[117] One informant noted that when lenders from out of town meet with
his broker colleagues, they do not ask, "How's the Chicago region doing?"
but rather, "What are my competitors offering in terms of interest rates?"[118]
City planners looked to developers during the Millennial Boom. "We were
dealing with a frenzy of permit and subsidy requests," with the immediacy
of the deal elevated over coordinated or future-oriented action. "We went
for projects that were 'shovel-ready' because they were less risky. I mean, the
developer was there and just waiting for us to say 'go.'"[119] Those projects that
might have been more beneficial to city residents but required protracted
negotiations or additional marketing were put on hold.

With little time for due diligence, mistakes were made: "We [the city] gave subsidies to developers who supposedly had firm commitments from tenants but these turned out to be only 'letters of interest. We might have known this if we'd had more time to investigate."[120] During construction booms, few actors undertake the detailed research and analysis necessary to detect changes in the overall market. And even if they are consulted, few data sources reflect of-the-moment trends. With professionals falling back on instinct and direct observation, warning signals about an impending slowdown are obscured by a cloud of collective enthusiasm, allowing markets to get overbuilt.

In periods of speedup, investors move their money out of older buildings and into the construction and acquisition of new buildings. The switching that occurs between old and new within the same locale—what can be considered, in Harvey's terms, a secondary "micro-circuit" of capital—is facilitated by the routinized practices of agents whose practices and judgments are highly localized. In order to see how real estate professionals become local agents of financialization and steer capital to new buildings in ways that encourage overbuilding (and further financialization), I turn now to the story of Chicago in the early twenty-first century.

PART TWO

Chicago in the 2000s

Downtown Chicago's Millennial Boom

The West Loop belongs to John Buck. A former broker from Texas, Buck began building in the area in 1981. The John Buck Company (JBC) went on to erect eight major office towers on Wacker Drive, the area's main thorough-fare and what became the new nucleus of Chicago's downtown office market, starting in the 1980s. Sleek but not showy, these "ribbons of glass" set the tone for boom-time development.[1]

JBC's work attracted imitators, setting off a rush to build that quickly out-paced local demand. By the early 2000s, vacancy rates were on the upswing, and rents were not registering the interest that brokers claimed existed for office space. Investor Sam Zell famously quipped in 2003 that "three idiots are building 4.5 million square feet of office space in Chicago."[2] Buck was one of them.

Buck's easy access to money led him to take risks. In 2006 his 155 North Wacker was only one-fifth leased—well below the standard threshold for construction financing—when JBC received an estimated $290 million in construction loans. The law firm Skadden Arps agreed to relocate from an-other Buck building at 333 West Wacker Drive to anchor the building, but initially no other tenants followed its lead. Nevertheless, once a ten-story building on the site was demolished, construction on Buck's forty-eight-story tower began in 2007.

Many of these risks paid off handsomely, helped along by the developer's good timing and reputation. Buck's UBS Tower at 1 North Wacker was de-livered in 2000 after JBC received construction financing from Bank One and Morgan Stanley. Well leased (over 90 percent occupied throughout the boom), it sold three times after delivery: in 2002 (to Deutsche Bank), 2008 (to Hines), and 2011 (to the Irvine Company). Over the course of these three transactions, the price of the building increased from roughly $300 to $450 per square foot. Another one of Buck's buildings, 111 South Wacker, cost $300 million to build in 2005 and in 2006 was sold to a German investment firm for about $410 million, or $410 a square foot—a record price for Chicago at the time.

Buck defended his decision to continue building in a time of oversupply. He noted how the office buildings erected just twenty years earlier (some of them by his own firm) had since become "obsolete."[3] In the 2000s corporate tenants wanted larger floor plans, fewer columns, and smarter infrastructure. Tenants dissatisfied with their 1980s buildings wanted to lease a more modern product. "As the market moves," Buck declared, "you must move with the market."[4]

Buck's statement implies that tenant needs and wants exist a priori and that they are what compel a developer like himself to build. Yet, as I have argued, the operation of financial markets and the routine work of real estate professionals also require a steady flow of new assets. While tenant demand is significant, it is produced in competitive social contexts and shaped through interactions with brokers, financiers, and building owners and managers. Buck, in fact, functioned as all three: JBC had a leasing and management division, it operated investment funds, and it developed buildings.

When he was a broker working on the original team that leased the Sears Tower in the 1970s, Buck knew that, in the absence of a precipitating factor such as a merger or major expansion, he would have to be partly responsible for germinating in tenants the desire to move. Fortunately, he and his brokers were some of the best in the business. Buck was able to convince law firms and corporate headquarters to leave the buildings they had occupied for twenty years to relocate to his new glass towers. During the Millennial Boom his buildings thrived. Accounting giant Deloitte & Touche and its main competitor, Ernst &Young, both relocated to Buck buildings across the street from each other on Wacker Drive. Large law firms such as Locke, Lord Bissell & Liddell, as well as smaller management consultants, traded in older premises for new buildings. Some, such as law firm Skadden Arps, relocated from one Buck-built tower to another, newer, one.[5]

Developers like Buck downplayed the extent of overbuilding because, from where they sat, the market did not look overbuilt. Their Class A buildings enjoyed high occupancy rates and record sale prices throughout the 2000s. Instead, it was the buildings from which tenants relocated that suffered the consequences. For example, Two Prudential Plaza (1990), in the East Loop, sustained high vacancies after Deloitte & Touche's relocation and by 2012 was on the verge of defaulting on its loans. The owner of the complex had purchased it during the acquisition frenzy of 2006 using commercial mortgage-backed securities (CMBSs) to finance the bulk of the purchase price. With crippling debt service payments, the landlord lacked the capital to stop the hemorrhaging of tenants to the West Loop. Such 1980s-vintage towers may not have been "obsolete" before the boom, but they became so in the wake of the overleveraged buyouts, relocations, and disinvestment that marked the boom and its aftermath.

The John Buck story illuminates the professional practices common to these pronounced bursts of building activity. Developers are emboldened by cheap debt, jet-fueled markets for acquisitions and securities, status-hungry tenants, high valuations, and the cheerleading of brokers and city governments to ignore demand trends and add to the fast-growing inventory of space.

Buck's 155 North Wacker was delivered in 2009, several years after the Chicago office market hit its saturation point. So was Donald Trump's International Hotel & Tower, a ninety-eight-story, $850-million structure that was the tallest skyscraper constructed in the United States in more than three decades. Economists attribute the seemingly irrational decision to continue to build in overbuilt markets to the lag time between getting the necessary approvals and financing and being able to physically construct a building. Time and space prevent supply from equilibrating to demand.

But even if Buck and Trump had possessed better information about the future, they still would have gone forward with their towers. Convinced of the uniqueness of their trademarked products, they were confident that tenants would emerge or upgrade when given the opportunity. Banks and investors eagerly wrote checks to them to underwrite construction. Charismatic local brokers and public and private financial incentives helped them cultivate an occupant pool for their products. Traditional market analyses and demographic trends could be discounted, as supply would create its own demand. The players in this game managed to convince themselves that square footage in a new Class A tower was indeed a coveted luxury good rather than a simple commodity.

Part II of this book discusses how the Millennial Boom played out in Chicago's commercial core. This first chapter is initially descriptive; it narrates the physical transformation of the downtown between 1998 and 2008, adding context and competitors to the Buck story. I then return to the question that animates the book: What caused the physical expansion experienced by Chicago and many other North American cities during the first decade of the twenty-first century?

The Historical Context of Downtown Development

Despite the extraordinary amount of construction activity in the first decade of the twenty-first century, the Millennial Boom was not the largest to occur in Chicago—or in other cities in the United States. There were four earlier periods of intensified commercial building during the twentieth century (fig. 4.1).[6] Each boom differed in terms of favored architectural styles, dominant land uses, and the motivations underlying the flurry of construction and demolition.

Each boom also differed in terms of the location of development. For the purposes of collecting spatial data, I define Chicago's downtown or central business district (CBD) as encompassing the area from Lake Michigan to Halsted on the west, and from North Avenue to Cermak on the south. This area is similar to what the City of Chicago refers to as the "Central Area" and is often broken down into submarkets, anchored by the Loop (fig. 4.2).[7]

Earlier booms set the spatial context for the expansions that followed them (fig. 4.3). The first two building booms, in the first half of the twentieth century, occurred after Chicago's reputation as a regional industrial powerhouse was firmly established. As such, they were characterized by periods of intense population and employment growth and technological innovation. Between 1860 and 1890, ten thousand new manufacturing and processing plants—primarily for steel, food, and clothing production—had located close to the city's railroads and waterways.[8] Chicago was dominated by a handful of large companies whose business operations originally took place on-site at their plants. Gradually, however, office functions migrated downtown. The Loop

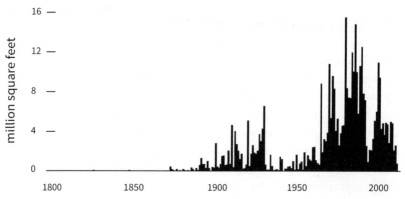

FIGURE 4.1. Office building deliveries in Chicago, 1800–2010. Data from CoStar Property Analytics.

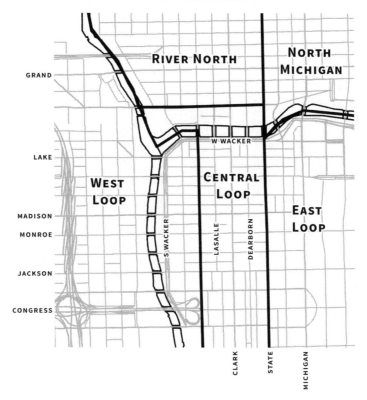

FIGURE 4.2. Downtown Chicago submarkets, 2000. Based on the CoStar Group's boundaries.

was also the retail hub for the region, anchored by large department stores, trading outlets such as wholesalers and futures exchanges, and service providers such as lawyers and accountants. It was the "central coordinating node for the rest of the metropolis."[9]

1901–1917

Approximately 191 major buildings were erected downtown during the first boom of the twentieth century.[10] These buildings reflected a new vertical expansion that was due to increased land values, the latest building technologies such as electric motors in elevators and steel framing, and the passage of a city ordinance in 1902 that doubled Chicago's building height limit to 260 feet (about twenty-two stories).[11] Many tall buildings were built in the "Chicago School" style; they were boxy structures with underlying steel frames clearly expressed through their grid-like facades that featured a "hollow-square" design with interior atriums and light wells above to provide light and ven-

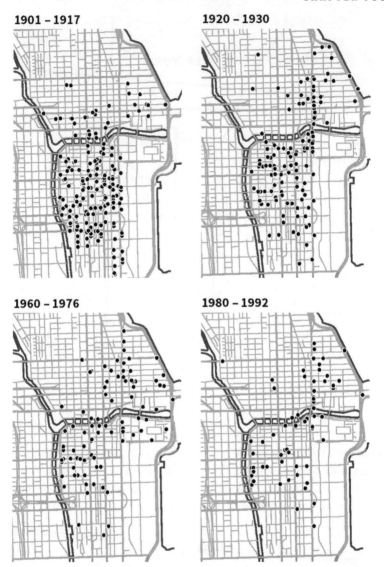

FIGURE 4.3. Locations of office building deliveries in downtown Chicago during twentieth-century construction booms. Data from CoStar Property Analytics; Frank Randall and John Randall, *History of the Development of Building Construction in Chicago* (Champaign: University of Illinois Press, 1999).

tilation to the surrounding offices.[12] Other buildings, influenced by the Columbian Exposition of 1893, incorporated an array of classical detailing and terra-cotta cladding.

The majority of buildings erected in the Loop during this time were intended for office use, although some incorporated banks, theaters, or retail outlets in their lower floors. Hotels and entertainment venues concentrated along Michigan Avenue, and a number of new department stores went up along State Street, which experienced phenomenal growth as the city's major retail district during this period. As retail uses spread out from State Street to Wabash, a number of new loft buildings were also erected in the Loop's wholesale area south of Monroe Street and west of Wells Street. The higher land values in the Loop forced some of the storerooms that were located on Wabash and Wells streets to move into the former residential area just south of the Loop, which began to develop as a warehouse center during this time. Likewise, warehouses and factories were established north of the Main Branch of the Chicago River. Many of these buildings, especially those erected for the printing industry in the South Loop, were constructed with heavy timber and reinforced concrete floors that could sustain the weight of the machinery housed within.

1920–1930

Chicago benefited from the economic expansion of the Roaring Twenties following World War I. Approximately 136 buildings were erected downtown between 1920 and 1930. Passage of the city's first zoning ordinance in 1923 allowed more usable floor space, and therefore more substantial towers, to be built above previous height limits, although allowable densities were still below those of New York City. The city also allowed for ornamental, unoccupied structures to extend far above the new height limits (e.g., the clock tower on the Wrigley Building). Buildings soared to heights of thirty-five to forty-nine stories. Many of these limestone Art Deco skyscrapers, often built as offices and hotels, were concentrated along LaSalle Street and Michigan Avenue. Public investments such as the widening of LaSalle Street, the new Michigan Avenue Bridge, and the double-decking of Wacker Drive facilitated a northward push (fig. 4.4).

Massive new buildings were constructed to accommodate banks experiencing growth from business operations and mergers. The Merchandise Mart, which encompassed two city blocks and drew smaller tenants away from the South Loop, was built speculatively, and other spec buildings were constructed on the northern portion of LaSalle Street. Widespread automobile use and the 1928 ban on curbside parking precipitated the construction of a number of high-rise garages in the Loop. Meanwhile, the warehousing and industrial functions of the South Loop spiraled into decline.

This boom was initially the result of pent-up demand (many developers had actually received permits in the 1910s but were unable to build because

FIGURE 4.4. Wacker Drive construction, 1925. Photograph courtesy of Chicago History Museum.

of World War I and strikes) and falling construction costs. Employment increased, as did Chicago's population, which grew 35 percent from 1918 to 1926. Bank employment in particular grew at a rapid clip as the city's specialization in this sector increased, and major banks moved to new quarters—sometimes just across the street. Land values in the city increased by 150 percent between 1918 and 1926.[13] Although employment growth began trailing off in 1927, the boom was extended through the use of new financial instruments that channeled capital into the construction of massive office and retail buildings. Guaranteed mortgage companies and security underwriting firms sold real estate bonds to the public just as corporations sold stocks. They used the funds to extend credit to building companies, the repayment of which they guaranteed to investors. Banks also provided "shoestring financing," which allowed debt to be taken out on 100 percent of a building's future value.[14] This boom left Chicago with a glut of downtown office space.

After the stock market crash, a wave of demolitions occurred in the Loop, clearing the area of many office buildings—some of them less than thirty years old—to make way for parking lots, garages, and one- and two-story "taxpayer" buildings whose rental income covered only the property taxes. Incredibly, almost a fifth of the buildings in the Loop (about 115 buildings) were torn down in the 1930s.[15] Most of these were older buildings whose high vacancy rates were due in part to the proliferation of competing stock built during the boom years. Chicago's Loop, Homer Hoyt wrote, "was once an area of new and vigorous growth in 1873 and an obsolete and blighted area in 1933."[16] Construction in the Loop came to a standstill, and little was built from the early 1930s until the mid-1950s.

1960–1976

Mayor Richard J. Daley presided over the long midcentury boom. It was attributed not just to post–World War II demand, but also to Daley's ability to make the downtown attractive to investors.[17] The first mention of the "city as regional center" that he envisioned is contained in the 1958 *Development Plan for the Central Area of Chicago.* The plan is decidedly postindustrial, with the downtown projected as a "command and control center for corporate and governmental institutions."[18] Indeed, by this time, much of the remaining light industrial and warehousing facilities had been pushed to the perimeter of downtown, making room for the fifty-story office towers and parking garages that were added to the mix of theaters, hotels, department stores, and public buildings. Mayor Daley was able to implement many of the plan's key elements—construction of McCormick Place, the University of Illinois at Chicago campus, federal buildings and plazas, numerous expressways and transit projects—as he was master of turning federal funds into local concrete. With the help of powerful advocates in Washington such as Congressman Dan Rostenkowski, Daley parlayed funds appropriated through Title I of the Housing Act of 1949 into several public sector-led efforts to stem middle-class suburbanization and white flight. He formed the Department of City Planning and the Public Building Commission to assist him with these efforts. And because plans were the key to accessing federal dollars, Daley planned: he directed his new agencies to create blueprints for future development, such as the *1966 Comprehensive Plan.*

The commercial construction boom, much welcomed after a long period of downtown dormancy, was slow to take hold after the Prudential Building—Chicago's tallest to date—went up in 1955. It was followed by sixty-four major buildings before the boom's end, adding about 50 percent more office space than what existed in 1960.[19] These buildings were mainly steel-and-glass towers built in the modernist International style. Helped along by more liberal interpretations of allowable densities, they were considerably larger in height and bulk than their predecessors. During this period "the Department of Planning had become completely aligned with the growth coalition agenda, facilitating the most permissive FAR and bonus system in the country."[20] The John Hancock and Amoco buildings were built during this time, as was the Sears Tower (completed in 1974), which was the tallest building in the world for twenty-five years. Institutional facilities such as the Federal Center complex and the Chicago Civic Center (now the Richard J. Daley Center) were also erected during this period, the latter featuring the largest all-welded structural skeleton up to that time. This period also saw the start of residential construction in the area north of the Chicago River, beginning with such projects as the famously honeycombed Marina City and Lake Point Tower to its east.

The shift from manufacturing to a service-based regional economy accelerated, and the number of white-collar workers in the Chicago region doubled. Meanwhile, the relocation of manufacturing to the suburbs also intensified during this period. Manufacturers had been moving from Chicago to the growing satellite cities of Elgin, Joliet, and Gary since the turn of the century, but the Interstate Highway Act of 1956 subsidized expressways that cut wide swaths through Chicago and its hinterlands, making automobile and truck transport more cost-effective for operators. Land developers sold lots on installment plans and erected new residential subdivisions in the suburbs. Businesses and households began to relocate farther away from the downtown. Industrial and warehousing facilities followed the I-55 and I-90 truck and rail freight routes west out of the city. Chrysler built its plant in exurban Belvidere in 1965, and intermodal terminals and distribution facilities sprung up in locations miles away from the city center.

Office uses also experienced a first wave of suburbanization in the 1960s. Secondary office markets were developed in the downtowns of older inner-ring suburbs such as Evanston as well as in edge cities such as Oak Brook and along the I-88 corridor. The other main office submarkets included those near Itasca, Woodfield, O'Hare Airport, the Edens Expressway, and Lake County. These employment subcenters became social and economic nuclei that competed with downtown. They offered tailored environments, larger swaths of rentable space, parking, security, and proximity to both suburban industrial facilities and the residential suburbs where white-collar employees increasingly lived.

1980–1992

Despite the election of Chicago's first non-machine mayor, Harold Washington, and the vicious "Council Wars" that stymied the implementation of Washington's progressive agenda, commercial real estate experienced a jolt of energy during the 1980s. Developers rebuilt the frame around the old core of the Loop, adding more than ninety major buildings and 30 million square feet of Class A office space.[21] New construction mainly took the form of office towers, fourteen of which were built along Wacker Drive, with another smaller concentration along LaSalle Street in the heart of the Loop. New hotels and three important public buildings—the State of Illinois Building (now the James R. Thompson Center), the Metcalf Federal Building, and the Harold Washington Library—were also constructed during this period. The cubic form of the International style was largely abandoned in favor of buildings with curving walls, sawtooth elevations and other kinds of distinctive tops, and richly decorated interiors.

A cluster of high-rise mixed-use buildings, including One Magnificent Mile (980 North Michigan), were built north of the Chicago River. To a lesser extent, the north end of the Central Area also featured new office buildings, hotels, and retail buildings. Meanwhile, the south end of the Loop was losing investment and consumers. The State Street Mall, a modernist pedestrian and transit corridor that opened in 1979, failed to stem the hemorrhaging of retailers to the North Loop and suburban malls.

The regional economy continued its transformation from one with a primarily industrial base to one highly dependent on producer services (finance, insurance, and real estate as well as law, management consulting, accounting, advertising, design, engineering, and architecture). For the city, 1979 was a watershed year when the number of people employed in services overtook the number employed in manufacturing for the first time. The office construction boom had its roots in this service employment growth, which increased the number of these jobs in the region from 3.7 million in 1979 to around 4.5 million in 1987.[22] Chicago was transformed from a regional hub serving local companies to a global financial and logistics center, with most of this activity concentrated downtown. Major corporate headquarters relocated to the Chicago region from other North American cities, and financial firms repopulated the central business district to be close to the trading and futures exchanges and management consultants.

But many of these same kinds of companies selected suburban office parks, which provided the downtown with dogged competition for office tenants. Moreover, despite the increase in service sector jobs, many of the new office buildings were built speculatively, with few tenant commitments. Developers bragged that "pre-leasing is for sissies" but found themselves shouldering the burden of a host of poorly occupied, "see-through" buildings.[23] The overbuilding of commercial real estate in the 1980s stands out even in such a cyclical industry; official office vacancy rates in downtown Chicago shot to over 20 percent in the early 1990s. Values and rents plummeted, and developers and lenders beat a retreat. The nail in the coffin was the 1992 relocation of Sears from its namesake tower to an office compound in suburban Hoffman Estates. The move added 1.5 million square feet of vacant Class A space to the downtown market.[24]

Residential development was a slightly different story. A turning point for the downtown's land use mix was the *Chicago 21* plan in 1973, which identified areas in the South Loop as potential residential "buffers" between the largely African-American State Street corridor and the historic Loop.[25] The first major project to enact the goals of this plan, Dearborn Park (completed in 1986), ushered in a wave of smaller industrial loft conversions in the South Loop, which eventually became the fashionable Printers Row neighborhood. Ambitious high-rise projects did not receive funding at the end of the decade because credit markets were already unraveling.

When Mayor Richard M. Daley took office at the tail end of 1980s boom, he expressed a strong desire to make the downtown a 24–7 destination. This dream required a diversity of land uses and, importantly, a substantial residential base downtown. It also required sufficient amenities to attract tourists and new residents, and to keep them spending on local retailers and services. Daley was able to accomplish these goals by riding the next construction wave.

The Millennial Boom, 1998–2008

By 2000 the Chicago region was one of two major office markets in the United States that were still core-dominated (the other being New York City). While most U.S. cities hosted the bulk of their office space outside of the CBD, almost 60 percent of the regional office space in the Chicago area was concentrated downtown.[26] Approximately a half million individuals worked there at the turn of the twenty-first century, with about 30 percent of all private sector employment in the city located in the downtown zip codes.[27]

The Millennial Boom, like most upticks in construction activity, occurred tentatively at first. The bust following the 1980s boom had been a punishing one, and only after a period of credit rationing and internal reorganization did banks resume lending. In the middle of the 1990s, commercial banks began cautiously providing construction loans to a series of retail projects north of the Chicago River in the already fashionable North Michigan Avenue submarket. These projects, in which a number of well-capitalized financial partners took the riskiest positions, were solidly pre-leased. In 1997 developers delivered the first phase of 300 East Randolph, a commission for insurer Blue Cross Blue Shield. The low vacancy rates inspired JBC to begin construction on One North Wacker Drive, the first speculative office building since the recession of 1990–1991. Other local developers took note and initiated a first wave of office construction, with deliveries occurring between 1999 and 2003–2004 (table 4.1).

The dot-com crash in the summer of 2000 and the ensuing recession

TABLE 4.1. Major Office Towers Delivered, Downtown Chicago, 1999–2009

	Address	Year built	Stories	Developer	Submarket
Union Tower	550 W. Van Buren	1999	17	Development Resources Inc.	West Loop
550 W. Washington	550 W. Washington	2000	19	Fifield Companies	West Loop
Congress Center	525 W. Van Buren	2000	16	Development Resources Inc.	West Loop
Dearborn Plaza/ Amalfi Hotel	20 W. Kinzie	2000	17	The Alter Group	River North
550 W. Jackson	550 W. Jackson	2001	18	Mark Goodman & Associates	West Loop
UBS Tower	1 N. Wacker	2001	50	The John Buck Company	West Loop
191 N. Wacker	191 N. Wacker	2002	37	Hines	West Loop
Quaker Tower	555 W. Monroe	2002	18	Fifield Companies	West Loop
ABN Amro Plaza	540 W. Madison	2003	30	Hines	West Loop
Citadel Center	131 S. Dearborn	2003	37	Prime Group Realty Trust	Central Loop
Chicago Transit Authority	567 W. Lake	2004	12	Fifield Companies	West Loop
Hyatt Center	71 S. Wacker	2005	48	Higgins Development and Pritzker Realty Group LP	West Loop
One Eleven S. Wacker	111 S. Wacker	2005	51	The John Buck Company	West Loop
One S. Dearborn	1 S. Dearborn	2005	40	Hines	Central Loop
Concord Place	550 W. Adams	2006	19	Fifield Companies	West Loop
Block 37	108 N. State	2008	16	Joseph Freed and Associates	Central Loop
One11 W. Illinois	111 W. Illinois	2008	10	The Alter Group	River North
155 N. Wacker	155 N. Wacker	2009	48	The John Buck Company	West Loop
300 N. LaSalle	300 N. LaSalle	2009	60	Hines	River North
353 N. Clark	353 N. Clark	2009	44	Mesirow	River North

Source: Data from CoStar COMPS; Crain's Chicago Business, "The next downtown office tower: One and done?," case study, February 25, 2013, http://www.chicagobusiness.com/realestate/20130418/ CRED03/130419770/

Note: Only those buildings ten stories or higher are included here. The three major buildings delivered in 2009 are included as part of the Millennial Boom (1998–2008) because construction on them began in early 2007.

stalled the build-out of several million square feet of office space that was in the pipeline. Many downtown office tenants downsized or vacated, which resulted in substantial "space givebacks." Moreover, concerns about tall buildings that arose after the terrorist attacks of September 11, 2001, put a dent in plans for several new skyscrapers.[28] New condo sales sputtered. Other plans went forward but were reconfigured. Developers of Kinzie Station, a residential complex northwest of the Loop, decided to build a second tower as rental units rather than condominiums.[29]

The recession officially ended in November 2001.[30] A second wave of office building began after that, and this one produced buildings that were, on average, larger, more architecturally ambitious, and sited in higher-profile locations than those built during the first wave. After reaching the pinnacle of the first development cycle in 2000, office building deliveries peaked again in 2005. Developers such as JBC (111 South Wacker), Hines (300 North LaSalle), Pritzker (Hyatt Center), and Fifield (550 West Adams) took the lead in building a new breed of exceptional office building that brokers labeled "Class A+" because of their size, location in the West Loop, modern interiors, and gross asking rents of over $50 per square foot.[31] Three of the largest skyscrapers—300 North LaSalle, 155 North Wacker, and 353 North Clark—started construction in 2007, coming online shortly after the bust in 2009 (fig. 4.5).

The boom produced twenty major office towers (see table 4.1). Each was over 200,000 square feet, and several, at over fifty stories, pushed past previous height limits. By 2009 about 20 million square feet of new office space had been added to Chicago's central business district.[32] This represented an almost 15 percent increase over the 134 million square feet that existed at the turn of the twenty-first century.[33] Although only two-thirds the size of the 1980s office boom (which added about 30 million square feet), the new boom gave a substantial boost to the city's overall supply. After the Millennial Boom, Chicago felt like a modern city architecturally; by 2010, over 50 percent of the downtown office stock had been built after 1980.

The skyscrapers of the Millennial Boom were built mostly in the contemporary steel-and-glass style and had a "classically-inspired architectural monumentality."[34] The buildings' facades tended to reflect classical forms and their building materials, rather than referring back to any specific moment in historical architecture (neoclassical, neo-Gothic).[35] In this sense, they were more modern than postmodern, especially when compared with the office buildings from the 1980s boom that riffed on the symbols of earlier periods (e.g., Philip Johnson's 190 South LaSalle, whose gabled roof recalled that of the demolished Masonic Temple). Chicago's skyscrapers did not gar-

FIGURE 4.5. 300 North LaSalle, 2012. Photograph courtesy of Scott McDonald, Hedrich Blessing.

ner the same aesthetic accolades as office buildings like the Hearst Tower or the New York Times Building in New York City. Critics considered them to be functional and understated, but not particularly inventive.

The speed at which these towers were erected was breathtaking. Developers describe this period as a time when they were "panicked," "crazy," and "pressured to make split-second decisions."[36] They raced against time, competing globally for materials—such as steel and drywall—that were in increasingly short supply. They tried to shave time off construction periods to avoid being saddled with higher interest rates. A representative of Hines noted that "when you borrow $200 million or $300 million, it's a heavy clock ticking even if you pay 6 percent rather than 8 percent. We're all at the mercy of interest rates."[37] Developers also sped up construction to ensure that they would tap the manic acquisitions market (the new office towers sold an average of 2.6 years after the date they were delivered).

A dramatic shift in land use occurred downtown during this period. Almost 75 percent of the new square footage added was residential, found in the high-rise condominium and multifamily apartment buildings, hotels, and dormitories concentrated in the East and South Loop (fig. 4.6). Local developers such as Magellan (Lakeshore East), Mesa (Heritage at Millennium Park), Fifield (Left Bank at K Station), Habitat (400 North LaSalle), Enterprise (Museum Park Towers and Lofts), Golub (the Streeter), and Terrapin (Burnham Pointe) were behind most of these projects.

The first residential projects of this boom hewed close to tested designs and locations. Conversions of old warehouses and industrial and office buildings had been taking place in the South Loop since the early 1980s. Even the new residential projects were somewhat derivative, many of them aspiring to an almost Victorian suburban ideal. For example, the 80-acre Central Station in the South Loop echoed the planning features of the previous boom's Dearborn Park. Both are inward-facing villages with small pocket parks and internal vehicular and pedestrian paths. A major expansion of Central Station was approved in 2000, and the master developers, Cleveland-based Forest City and Chicago-based Fogelson, subsequently allowed other builders to develop additional vertical towers of condominiums and rental units there (fig. 4.7).[38] Perhaps the most significant validation of the area's cachet was when Mayor Daley himself moved from his childhood neighborhood of Bridgeport to a townhouse in Central Station.

Developers added about 34,000 new housing units downtown between 2000 and 2010, with about three-quarters of these in larger buildings of twenty-five units or more.[39] This resulted in a 50 percent increase over the 68,000 units that existed in 2000 (not including demolitions). High-rise condo construction in the East and South Loops did not begin in force until the latter part of the boom, accelerating in the years 2005 and 2006 and then again in 2009.

The Millennial Boom had the spatial effect of extending the traditional boundaries of the Loop into new submarkets.[40] Zoning and historic preservation regulations permitted redevelopment of all but a handful of areas— such as the majestic Michigan Avenue streetwall, which was protected by a landmark designation and a preservation ordinance. The submarkets on the

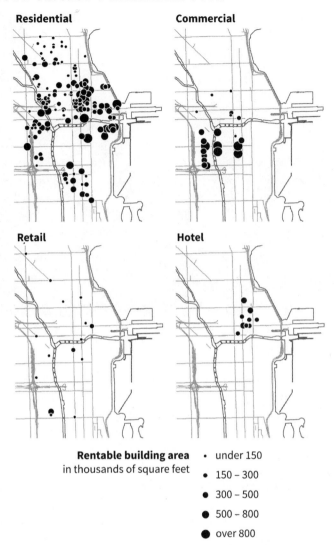

FIGURE 4.6. New construction in downtown Chicago by land use, 1998–2008. Data from CoStar Property Analytics; Goodman Williams Group downtown database.

fringe of the historic core were initially less built up, with available vacant land, smaller building footprints, and shorter towers. Before the boom, these submarkets were cut off from the Loop by a "value moat" of rail lines, aging building stock, and physical impediments, such as the Chicago River. Land prices and rents were lower in these areas than in the rest of the downtown, but as explained in the following chapter, developers there received generous assistance from the municipal government, including subsidies to deck

FIGURE 4.7. Central Station development. Photograph courtesy of Steve Hall, Hedrich Blessing.

over rail lines. As in Lower Manhattan in the 1960s, a combination of public investment, private interest, and brokered preferences helped to turn these "new" submarkets into sites of prestige and status. Doing so also involved denigrating the spaces left behind—found mainly in the historic core and along Michigan Avenue—as "blighted" and "obsolete," difficult to retrofit, and in need of public assistance.

The bulk of the new office towers were constructed along South Wacker Drive in the West Loop and in the River North submarket. Office development pushed past the Chicago River to the Kennedy Expressway, as more than ten of the major office towers completed during the decade were built west of LaSalle Street. Investors were willing to pay a premium for these newly desirable addresses, particularly for the office buildings east of the river along Wacker Drive. On the west side of the river, land was less expensive and buildings smaller, often with a single tenant occupying most or all of the space. Both sections benefited from their proximity to the Ogilvie commuter rail station and to the interstate. By the end of the boom, more than 60 percent of the total Class A office space in the downtown was located in the West Loop submarket.[41] Some new retailers and hotels followed suit, although the majority remained concentrated in retrofitted historic structures near the Loop's epicenter at State and Madison. National chain retailers also set up shop in modified big boxes along the Loop's southern border on Roosevelt Road.

Residential development similarly pushed the traditional borders of the downtown outward, although its geography was distinct from that of the office towers. Since the 1980s small-scale conversions had taken place in the spaces left behind by manufacturing and office tenants in the South Loop and Near North Side.[42] In the early phases of the boom, new lower-density townhouses arose on the fringes of the downtown, such as University Village adjacent to the University of Illinois at Chicago campus. When high-rise apartments began appearing en masse, they were built in different areas: namely, the East Loop (such as Lakeshore East, Chicago-based Magellan Development's master-planned community between Millennium Park and the Chicago River) and the South Loop (anchored by the Central Station). Some residential developments eventually penetrated the historic office core, such as the Heritage (delivered in 2004), located a half block west of Millennium Park. It adhered to historic preservation requirements by restoring four architecturally significant facades along Wabash and building a two-tiered tower fifty-seven stories above them. Building heights soared beyond anything experienced previously, adding variety to the skyline while breaking up the streetwall with setbacks and landscaping.

The 2005–2006 period marked a transition as condo buyers began pulling back. By the end of 2007, about 7,700 new and rehabbed condos and townhouses downtown sat unsold.[43] To make matters worse, the buildings added to the market during the second wave were substantially larger in size and their apartments were priced higher than most of the units for sale early on. The undulating eighty-two-story Aqua (fig. 4.8) came on line in 2009, by which time the condo market had ground to a halt. The office market's timing was not much better: the Alter Group's One11 West Illinois, JBC's 155 North Wacker, and Joseph Freed's Block 37 mixed-use tower were delivered just in time for the credit crisis to burst the cycle of easy money and asset appreciation.

Developers were aware that they were contributing more stock to an overbuilt market and that their competitors were completing construction with unsold inventory. They substituted hotels for condos as the hospitality industry took off after years of being unable to keep up with the increasing demand from tourists, trade shows, and conventions. Thirteen new hotels were constructed during the Millennial Boom, adding about 4,200 rooms to the market. When hotels started to lose their luster, developers converted their projects to rental buildings or student housing—a growing but price-limited market with sixteen college campuses downtown. Approximately 5,800 student units were added between 2000 and 2010.[44]

FIGURE 4.8. The Aqua, 2013. Photograph courtesy of Kevin Dickert.

But even these changes in building type and use could not stave off the inevitable end of the party. Contrary to the popular image of busts, the cranes did not stop in their tracks in 2007, nor did the banks repossess the skyscrapers whose owners were delinquent in repaying their loans. *Crain's Chicago Business* published a series on the housing market entitled "After the Boom" on October 30, 2006, but developers were still completing ambitious office towers in the final months of 2009, two years after what economists identified as the beginning of the recession.[45] By that time half-completed, foreclosed properties had become an all too common feature of the landscape. For those in affiliated real estate fields, it had been a good run, but in the words of one banker, "we all knew it wouldn't last forever."[46]

What Caused the Millennial Boom?

The rapid pace of construction during booms makes it difficult to pause long enough to single out the factors propelling that activity. The demand- and supply-side catalysts blend into one another as the growth machine fires on all cylinders. During the Millennial Boom the City of Chicago granted building permits, corporate management considered relocation plans with their real estate teams, banks made loans, aldermen altered zoning and height restrictions, investors raised equity, brokers signed on new tenants, valuations peaked, and new developers entered the market. No one asked *why* the activity was happening, just *how* they could take advantage of it. Optimism prevailed—before the momentum dissipated and accusing fingers were pointed for ignoring the telltale sign of the bust.

Building booms traditionally have been explained by factors governing either the demand for or supply of space. In the popular demand-side account popularized by neoclassical economics, "property is treated as a derived demand . . . buildings are, in effect, deemed to exist or appear in response to the needs of industry."[47] In contrast, the Millennial Boom in Chicago was only weakly demand-driven. This demand emanated mainly from existing office tenants and households relocating from older buildings rather than from in-migrants to the region. Moreover, despite advances in building technologies, existing tenants were not clamoring for the new offices; the modern skyscrapers did not promise sizable enough differences in operating costs or productivity to warrant mass relocation.

More compelling explanations for the surge of new commercial construction that occurred during this ten-year period can be found in the changes taking place in global financial markets and in the professional practices that validated the rapid intra-urban movement of capital and tenants. But before describing the influence of these supply-side factors in the remaining chapters, it is worth scrutinizing the occupant markets in Chicago during the Millennial Boom for new growth and changing preferences.

DEMAND FOR NEW OFFICE SPACE

When the Millennial Boom began in the late 1990s, the Chicago region had hit its stride as a postindustrial economy. It was well established as a location for corporate headquarters and firms in professional services, health care, and transportation. Approximately 185 publicly traded companies with more than five hundred employees were headquartered in the Chicago region in

2000 (making it fourth in the country after New York City, Los Angeles, and San Francisco).[48]

As the boom progressed, a number of businesses moved their headquarters to Chicago from outside the region. The most high-profile headquarters relocation was that of aerospace giant Boeing, which moved from Seattle in 2001 after considering both Dallas and Denver. Boeing had been in the Puget Sound area since 1916, and its corporate culture—not to mention the thousands of unionized workers that it employed—was highly embedded in that location. The decision to relocate to Chicago was perceived as a way to distance management from past production methods and to embrace a future that emphasized shareholder value and financial engineering practices.[49] Chicago was selected not only for its location in the middle of the country (and close to newly acquired McDonnell Douglas in St. Louis), but also because of the perception of the stability and financial connectedness of the city.

Regardless of its motives, the fact that Boeing selected Chicago symbolized the city's economic resurgence and cemented its position as a global powerhouse. Other firms with an international reach followed suit. Both Mittal Steel, an Indian company, and Veolia, the world's largest environmental services company, selected Chicago for their new North American headquarters.[50] In 2008 MillerCoors signed a fifteen-year lease for space on Wacker Drive, relocating approximately three hundred employees from Wisconsin and Colorado. Purveyor of spatial data Navteq and United Airlines moved their headquarters downtown from Chicago suburbs. New start-up businesses, such as CareerBuilder and Infinium Capital Management, also constituted some of the demand for office space.

These instances of new business starts and corporations moving to Chicago from other parts of the region or globe, however, were the much-touted exception rather than the norm. In fact, for every major firm that relocated from outside the city, an existing business moved away or was downsized after a merger or acquisition. For example, when New York–based JPMorgan Chase bought and closed Bank One in 2004, it put approximately 2 million square feet of space back into the Loop market. Two years earlier, Chicago lost the corporate headquarters of accounting firm Arthur Andersen, which broke up following the Enron scandal. The John Hancock Center's largest office tenant, insurance brokerage Near North National Group, collapsed in 2003. Montgomery Ward, Unicare, Ticketmaster, GE Capital, ABN Amro, Altheimer and Gray, Bank of America, and Accenture were all acquired and downsized, closed, or left the city. Heavy layoffs at Motorola, Tellabs, SBC Communications, and United Airlines also stymied growth during the first decade of the new century.

More revealing than individual cases of closure and exit are the aggregate employment figures for the City of Chicago. Annualized growth rates for private employment in Chicago's CBD slowed to an average of just 0.0003 percent between 1998 and 2009.[51] When only those sectors in which the majority of employees work in office buildings are considered, the rates of change are slightly better (0.006 percent), but they are still far below the 4 percent rates found in the United States in the previous construction boom of the 1980s—a boom that, despite the underlying job growth, still left millions of square feet vacant in its wake.

Chicago's downtown employment increased gradually starting in 1993 and peaked in 2000. However, the recession of 2001 had a severe impact on the entire region, which lost more than 45,000 white-collar jobs between 2001 and 2003.[52] Job losses in Chicago exceeded those in New York, even though that city was a target of the September 11 attacks. By 2005 the Chicago region had replaced only about 4,500, or 10 percent, of the jobs lost during the recession. The failure of many dot-com start-ups and layoffs at old-economy firms left millions of square feet of subleasable space downtown.[53]

Because of the slow job growth, new office deliveries outpaced the rate of downtown employment change, as figure 4.9 shows starkly. According to dominant theories in urban economics, supply should be tracking demand and following a similar, but slightly lagged, trajectory over time. In Chicago, however, while office-user employment and total employment rose slightly and then dipped back to the pre-boom rates, the rentable building area downtown charged upward. The total inventory of office space grew by 15 percent, yet by the end of the boom there were approximately 31,000, or 8.3 percent, fewer office-using jobs.[54] Once-dominant industries such as finance and insurance accounted for a shrinking share of total leasing activity after 2001 (table 4.2). The discrepancy between the rates of change in supply and demand shows that they had become decoupled as the boom proceeded—either because too many buildings were erected or because demand fell off unexpectedly.

The city's weak growth rates were partly due to the slowing of the national economy, which eventually culminated in the recession of 2007–2009. Pundits called the period after the shorter 2001 recession a "jobless" recovery, one that improved profit prospects for corporations and investors but created no incentive for increased employment.[55] But even during a decade when most parts of the country were struggling, Chicago's employment losses were the worst among the country's ten largest metro areas.[56] The city's unemployment rates remained two percentage points higher than the U.S. national average for most of the boom.

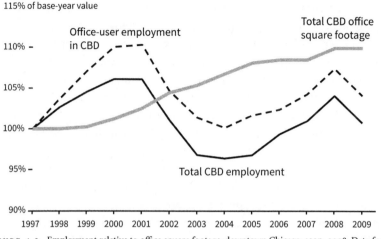

FIGURE 4.9. Employment relative to office square footage, downtown Chicago, 1997–2008. Data from Illinois Department of Employment Security, *Where Workers Work*, 2010; CoStar Property Analytics.

During the prior boom of the 1980s, the shift from manufacturing to services and the surge of women and baby boomers entering the workforce created the uptick in FIRE employment that helped to fill the new offices. The Millennial Boom can claim no such underlying demographic or economic trend to support its office building expansion. During the 2000s the office worker population aged and retired.[57] And even those tenants that were expanding their workforces were paring down their space requirements. Corporate tenants continued to reduce the amount of space rented per employee as telecommuting, outsourcing to offshore and off-site locations, flexible cubicles, and open office arrangements became more common.[58] Even "space hogs" such as law firms scaled back.[59]

But just how much of the office space downtown should be considered "overhang" at the end of the boom? Published vacancy rates hovered between 12 to 14 percent, peaking at just under 18 percent in 2005 after several towers were delivered (fig. 4.10).[60] Taken on their own, these rates tell us little; lacking benchmarks that could reflect Chicago's "natural" vacancy rates, we do not know if they are exceptionally high or low or how they were measured in the first place (for example, whether sublet space is included). Moreover, it may be possible for demand to quickly fill what appears on the surface to be oversupply if developers refrain from adding more stock.[61]

To give some indication of the degree to which demand and supply were balanced during the boom, I estimate the utilization of downtown offices using data on total employment and the average square footage occupied per employee. The estimate for average annual office employment among private

TABLE 4.2. Private Employment by Sector, Downtown Chicago 2001–2008

	2001	2002	2003	2004	2005	2006	2007	2008	Change, 2001–2008 (%)
Finance and insurance	89,403	88,662	87,345	87,435	86,389	85,098	83,576	82,542	−7.7
Professional, scientific, and tech services	89,293	82,645	76,140	74,825	76,742	79,822	84,283	88,991	−0.3
Administrative, support, and remedial services	36,237	31,997	30,385	30,249	31,203	32,887	32,143	34,406	−5.1
Information	18,100	17,287	15,341	14,550	14,521	13,552	14,384	13,930	−23.0
Health care	15,155	16,813	15,859	16,507	16,892	17,594	18,052	19,240	27.0
Retail trade	13,703	12,410	10,321	10,167	8,092	8,595	7,656	8,061	−41.2
Accommodations and food services	13,210	12,695	12,565	13,147	12,907	13,550	13,531	13,919	5.4
Other services	10,489	11,543	11,704	12,272	11,957	12,542	12,723	13,187	25.7
Real estate	9,246	9,489	8,491	8,484	8,360	9,218	9,633	9,174	−0.8
Mgmt of companies and enterprises	6,418	6,906	8,331	8,811	8,768	9,243	9,418	10,193	58.8
Manufacturing	5,830	4,859	3,434	3,184	3,041	2,608	2,453	2,518	−56.8
Arts, entertainment, recreation	5,590	5,480	5,307	5,201	4,505	5,724	5,498	5,819	4.1
Wholesale trade	5,541	4,559	4,755	4,105	3,971	5,264	5,227	5,226	−5.7
Education services	5,288	6,106	5,984	7,112	7,033	6,737	6,960	6,987	32.1
Utilities	3,918	3,336	2,964	2,902	2,362	2,479	2,443	1,784	−54.5
Construction	3,519	3,488	3,474	3,173	3,233	2,938	3,219	3,439	−2.3
Total private employment	335,480	321,797	305,229	304,702	302,417	310,549	313,359	320,721	−4.4

Source: Data from Illinois Department of Employment Security, *Where Workers Work* (2010).

Note: Categories with fewer than 1,000 workers do not appear, although they are included in "Total private employment."

companies in Chicago's Central Area for 1998–2008 is 353,487 individuals.[62] Other studies estimate that approximately 10 percent of total Loop employees worked in the public sector, and if this assumption applies to the entire CBD, 39,276 government employees worked there—for a total of 392,763 of public and private employees.[63]

If each of these office workers required 280 square feet of gross space, then

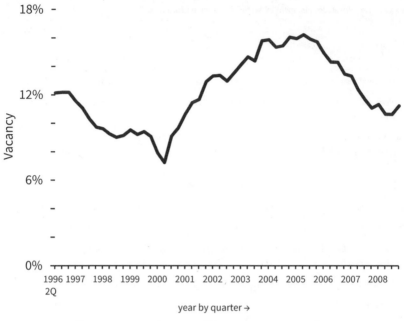

FIGURE 4.10. Office vacancy rates, downtown Chicago, 1996–2008. Data from CoStar Property Analytics.

the total demand for office space should be 109,973,733 square feet.[64] This conservative estimate suggests that the approximately 154 million square feet of rentable office space available in 2009 was about *44 million square feet* too much.[65] Although experts argue that a 5 to 10 percent vacancy rate is desirable so that businesses can move or expand, having no or little market demand for almost *one-third* of the building stock at the end of the building boom suggests a deeper problem.

Yet published vacancy rates never reached this 30 percent mark. One of the reasons for the discrepancy is the plethora of what brokers call "shadow space": inventory currently leased by tenants that is not fully utilized or occupied by them. Tenants typically shed excess space by subletting to other tenants or by negotiating contractions directly with landlords. Published rates are "direct" vacancy rates that do not count space being marketed for sublease. Moreover, landlords and their representatives go to great lengths to prevent information about shadow space and vacancies from entering the public domain, as such information lowers the price they can negotiate from future tenants or buyers. Some boom-time observers went so far as to condemn published vacancy rates as "irrelevant," given the discrepancies between the amount of space listed as vacant and the amount being marketed as available. As job losses started mounting after the financial collapse, one observer noted, "At no other time in the past 25 years has there been so much unlisted, and therefore undetected, space available for lease."[66]

Like vacancy rates, absorption rates reflect the extent to which demand and supply are in sync. They measure the net change in occupied space over a period of time (in most cases, between the beginning and end of a quarter). If absorption rates for a period are close to zero, then one might infer that growth was not occurring in the market, but rather that existing businesses occupied the new square footage, releasing just as much vacant space in the move—in other words, that there was no net gain of occupied space. Negative absorption may indicate either a surplus of space left behind after internal relocations or the possibility of tenant out-migration.

Downtown Chicago absorption rates reveal, in the aggregate, net stasis.[67] A report on the downtown market from real estate firm Grubb and Ellis states, "Although pre-leasing for the new office towers was high, tenants were being drawn from existing properties."[68] The market registered more negative net absorption rates during the Millennial Boom than it did positive ones. The statistics improved by 2006, but some of the absorption was "fake" in that square footage was taken off the market not because it was occupied, but because it was converted into commercial or residential condominiums. And even this small upward trend reversed itself in 2008 as the recession forced tenants to downsize and deliveries late in the boom pushed total supply to the brink.

Tenant data confirm the negative absorption statistics measuring building occupancy. In the Chicago metro region, 4,899 FIRE establishments moved locally at least once in each year between 1999 and 2003, for an average of 5 percent of all establishments.[69] Applying this metric to the downtown, we can impute that 50 percent of all FIRE establishments moved at least once during the decade-long building boom. Of those establishments moving at least once between 1999 and 2003, 52 percent remained within the same five-digit zip code, with only the four-digit locator code of their zip code changing. In many cases in the downtown, zip codes cover mail carrier routes that are only a few blocks long and wide. As such, a move often indicated a change of address within the same office tower or just down the street from a business's previous location.

Combined, these data suggest that growth was not the primary motivation for new construction. Gains from corporate entrants and expansions were cancelled out by job losses and out-migration, which left significant excess capacity in their wake. The building boom was either a cause or effect of a series of internal tenant moves that left about as much, if not more, space on the market as they occupied.

DEMAND FOR RESIDENTIAL SPACE AND OTHER USES

Rather than grouse about office overbuilding, boosters focused on the return of residents as well as retail, entertainment, tourism, and educational uses to the downtown after a long hiatus. Growth there looked like it was stimulated by renewed interest in the area as a place to live, learn, visit, play, and shop. Many non-office commercial uses appeared during this period, including rehabbed retail outlets on State Street to rival the more upscale North Michigan Avenue corridor.[70] But employment figures still show that by the boom's conclusion, retail jobs had declined and the office sector actually constituted a *larger* share of all downtown employment than it did when the boom started.[71]

The main source of underlying demand for real estate downtown was not new or growing businesses, but households, investors, tourists, and students purchasing or renting new residential units. Residential population downtown grew by 41 percent from 2000 to 148,133 individuals in 2010 (table 4.3),[72] by which time over 5 percent of city residents lived downtown.

The residential rediscovery of downtown was helped along by earlier rounds of gentrification that brought affluent households, students, and white-collar workers to the area. The interdependence between the office core and nearby prosperous neighborhoods created synergies as the CBD provided proximity to workplaces, lifestyle-enhancing amenities such as restaurants and cafés, and the density and street life sought by residents. This

TABLE 4.3. Population Change in Downtown Chicago, City of Chicago, and Chicago Metropolitan Statistical Area 1950–2010

	Chicago Downtown		City of Chicago		Chicago MSA	
	Population	Change (%)	Population	Change (%)	Population	Change (%)
1950	134,553	—	3,620,962	—	5,761,484	—
1960	102,612	−23.7	3,550,404	−1.9	7,017,024	21.8
1970	85,740	−16.4	3,366,957	−5.2	7,886,829	12.4
1980	81,032	−5.5	3,005,061	−10.7	8,052,932	2.1
1990	84,181	3.9	2,783,726	−7.4	8,182,076	1.6
2000	105,336	25.1	2,896,016	4.0	9,098,316	11.2
2010	148,133	40.6	2,695,598	−6.9	9,461,105	4.0

Source: Data from U.S. Bureau of the Census, "Population and Housing Tables," 1950–2010, http://cps .ipums.org/cps.

spatial restructuring reflects a changed thinking about the geography of the city. Downtown office markets across the country have been losing ground to residential development as well as other entertainment and institutional uses.[73] Chief among these institutional uses in Chicago is education: by 2010, 65,000 students attended classes in the Loop—more students than at any one Big Ten university.[74] Another bright spot has been tourism. Approximately 32 million leisure travelers from across the country visited Chicago in 2008, and travel volumes increased by 42 percent between 2000 and 2008.[75] The "tourist bubble" fed off the amenities catering to the downtown office workers and suburban commuters, sharing many of the same services and drawing energy from the same density.[76]

The transformation was also aided by the public planning initiatives discussed in more detail in the following chapters. The 2004 zoning code rewrite allowed for more flexibility of land uses while protecting the traditional core of office buildings. The city's use of tax increment financing funds to rehab historic facades, to underwrite retail development, new public schools, and dormitory construction, and to modernize transit stations also encouraged residential growth. But many argue that the most significant factor was the building of Millennium Park—Chicago's answer to New York City's Central Park and an urban playground of public art, landscaped gardens, and entertainment options—on 25 acres of underutilized space abutting the lake.[77] This public-private undertaking raised both nearby property values and the profile of downtown living.[78]

Although the new residential uses added vibrancy and vitality downtown, this market too became decoupled from market demand and was eventually overbuilt. The mismatch is evident in U.S. Census data, which tracks new housing units as well as household formation (a better measure of housing demand than aggregate population statistics). In the census tracts covering the downtown, the number of households increased by about 22,800 between 2000 and 2010, while the number of housing units increased by 34,000 during the same period.[79] Thus, for every new household in downtown Chicago, almost a housing unit and a half was added.

By the end of 2008, almost 5,500 condos sat mothballed.[80] The rental apartment market remained strong. But stalled development projects were visible, particularly in the South Loop, where the surfeit of unsold inventory led reporters to rename it a "dead zone."[81]

That the decline in this market coincided with the recession and the loss of tax breaks for new homeowners still could not obscure a structural problem for the downtown market: the population of the City of Chicago and of Cook County as a whole was decreasing. Even though it fared better than its

Rust Belt neighbors, Chicago declined in population by 7 percent between 2000 and 2010 (see Table 4.3). Approximately 200,000 Chicagoans left for other parts of the region and the country. These losses undercut the gains of the 1980s and 1990s, when the city added over 112,000 people, many of them foreign-born.[82]

The discrepancy between rates of population change in the city as a whole and in the smaller microcosm of downtown suggests that the new residents were not really new to the region but, like office tenants, were relocating from other parts of it. Household relocations in Chicago increased from 14.5 percent of all households in 1998 to 18.3 percent in 2007.[83] More than 70 percent of the movers in Chicago in 2007 originated from other parts of the city and Cook County—as opposed to the rest of the state, other parts of the United States, or abroad.[84] Market analyses, such as an Urban Land Institute case study of the Palmolive Building, found that many of the buyers of new residential units were moving from neighboring high-end areas such as the Gold Coast.[85] They also came from the suburbs as part of a much publicized "return to the city" movement associated with baby boomer retirees and empty nesters able to cash out their equity from their previous residences or purchase a second home.

During the 2000s white, affluent households increased their presence downtown and on the lakefront, while their numbers decreased on the city's west and northern fringes.[86] Urban-minded young creative types, college students, and white-collar workers seeking to reduce their commute times flocked to the new developments as interest rates on mortgages plummeted. Although overall inflation-adjusted household incomes were stagnant in Chicago, those for the 95th percentile, the intended market for the new high-end condominiums, increased 46 percent from 2002 to 2008.[87] Inner-city affluence led many to declare that Chicago was in the middle of a demographic inversion, with low-income people of color on the fringes of the city and the wealthy living in high-rises downtown.[88]

DEMAND FOR HIGHER-PERFORMING BUILDINGS

Even if occupant demand was not growing, it is possible that it was changing. Was the building boom of the 2000s motivated by a desire for more efficient buildings that could satisfy changing needs?

Several of the new office towers—particularly late deliveries such as 111 South Wacker and 300 North LaSalle—gained renown for their innovative interior layouts, environmentally friendly building systems, and scale. Like other contemporary skyscrapers, they were engineered to rely less on heavy masonry facades and interior partitions than on their structural frames. As a result, the new buildings could offer more unobstructed space and natural light. They were better insulated thanks to glass curtain walls and more sturdy and wind-resistant. Nine new office towers received Leadership in Energy and Environmental Design (LEED) certification, meaning that their design incorporated the latest advances in water and energy efficiency, indoor air quality, waste management, and sustainable materials.

The adoption of such building innovations is not easy to attribute to either producer self-interest or user needs; they result from negotiations between tenants and developers, moderated by brokers, architects, engineers, and building managers.[89] Some of the more savvy tenants reached out to developers, expressing an interest in specific layouts and building amenities. Before the law firm Kirkland & Ellis relocated its Chicago offices to 300 North LaSalle, it worked with the developer, Hines, to seek LEED certification.[90] Tenants taking a strong interest in the efficient workings of building systems was a change from previous era. According to one broker, in the 1980s tenants sought "views and glitz and private washrooms for partners and gold leaf in the lobbies . . . I used to sell 'features' of buildings to a bunch of CFOs who had no real estate experience. Now I sell 'benefits' to seasoned real estate professionals."[91]

Specifically, tenants wanted buildings with more exterior offices, wider core-to-window spans, more open workspaces, and large blocks of contiguous space uninterrupted by columns or elevators. Noted one architect who worked with Chicago law firms:

> You see that buildings that have been built in Chicago in the last cycle have followed that truth . . . Both [projects] that [law firm] Jenner was looking at had exceptionally smart layouts for the geometry of the building . . . things like a 45-foot clear span from window to core, no interior columns along that main span, long regular compositions of facades, a nice regular column grid that allows for reliability in partner and associate office layouts, and the flexibility of interchanging those two modules . . . Newer generations of buildings are the only ones that law firms, especially at the scale of Jenner, Kirkland and Baker [other law firms], can really afford to even consider.[92]

Tenants also wanted "smarter" infrastructure, such as wiring built into the walls and floors, cutting-edge security, and automated systems that respond to data collected by sensors instead of human decisions. One partner serving on his law firm's real estate committee noted that its move from LaSalle Street to a new building on Wacker Drive was based on the need for more wired and flexible space: "Our previous building [built in 1987] was completed before the tech revolution really occurred. Now all the wires need to be under the floor, we don't need a law library of the size we used to pay for, we don't need space for secretarial pools any more . . . Also, we stopped meeting in individual offices and so we needed fewer 'ego' offices and more shared space."[93] A facilities manager at another major law firm added that while the costs associated with moving would be greater, "we would get a brand new space that we could really control."[94]

Tenants were also interested in new locations because geography could produce material upgrades in quality. This was particularly true for the new residents of the downtown condo and apartment buildings, who saw the downtown as providing economic and lifestyle advantages over suburban locations and more far-flung city neighborhoods. For office tenants, a West Loop address meant better sight lines from the perimeter of the built-up core as well as proximity to amenities such as Ogilvie Station, where the commuter rail system dropped off white-collar employees coming in from the North Shore.

Demand for these amenities should be reflected in the price tenants are willing to pay. And rational tenants will pay more for new buildings only if they do better in them: they need to either be more productive or save on expenses so that the costs to occupy space consume a similar or smaller share of their overall budgets.[95]

Market feedback during the Millennial Boom confirms the idea that major tenants were willing to pay more to lease new space. The rental premium for new Class A space (as opposed to Class A buildings constructed before 1998) was over $4 per square foot.[96] The popular West Loop, home to many of the new Class A towers, consistently posted lease rates about $2 per square foot higher than the rest of the downtown. Between 2006 and 2009, asking rents grew faster in Class A buildings than they did in Class B ones (17 percent compared with 11 percent). In Chicago as well as nationally, new LEED-certified buildings commanded higher rents, assessed values, and market prices than their non-LEED (and likely older) counterparts.[97] While vacancy rates increased throughout the boom for all classes of buildings, new Class A buildings had, on average, comparatively better occupancy than older Class A or B and C buildings throughout the 2000s (fig. 4.11).

It is not clear, however, that tenants paid higher rents for new buildings because they offered performance advantages over existing buildings. Other evidence suggests that movement within the downtown did not always result in significant increases in revenues or decreases in expenses. In fact, as I will describe in more detail in chapter 6, the most common trend during this

Class A

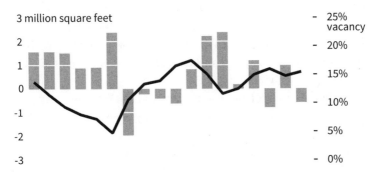

Class B & C

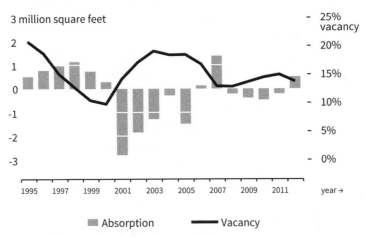

■ Absorption   ——Vacancy

FIGURE 4.11. Vacancy rates and absorption for Class A and Class B & C office buildings, downtown Chicago, 1995–2012. Data from CoStar Property Analytics.

period was for relocating tenants to trade in slightly older for more expensive contemporary construction with a modest technological edge.

While new buildings are intended to be more energy efficient than older stock, they can also be more expensive to operate. Several indicators of quality that the new buildings advertised—including better lighting, elevator service, aesthetic appeal, and ambient comfort—actually require higher short-term utility and maintenance costs to produce. One Chicago broker who had relocated many law firms to new buildings mused, "They [tenants] are not really saving yet. Many of them shrunk their head count but they're still paying for a lot of unused space."[98] Studies suggest that the savings from buildings officially designated as "green" may not materialize at all or may not show up for some time.[99] One tenant of a new LEED building claimed, "LEED was not that important for us. It was more for show. Our real estate group said 'as long as it doesn't cost us more' we're OK with it."[100] These may be some of the reasons why data on operating expenses for office buildings in downtown Chicago show them increasing at a faster rate than building incomes did between 1998 and 2010.[101]

The symbolic value of these amenities and designations may outweigh their material benefits. This could explain why even if the utility cost savings of new buildings materialize and are capitalized into higher rents, they still cannot fully account for the rent and sale price premiums paid to lease and purchase new Class A buildings. For example, one national study found a 13 percent sale price difference between LEED-certified office towers and comparable buildings, yet new Class A office buildings in Chicago sold for over 50 percent more than slightly older ones during the boom.[102]

Similarly, the productivity gains associated with new buildings—such as the claim that new buildings help tenants to attract top talent—are hard to pin down. Productivity trends for office-using businesses nationally were actually negative during the boom years; many service sector industries, including law firms, saw their output rates decrease.[103] Revenues and profits per partner for Chicago law firms, many of which had moved to new spaces, dropped precipitously between 2007 and 2008 and continued to decline as the recession wore on.[104]

If new buildings did help tenants become more productive, these gains may not have been large enough to compensate for the greater rent burden those tenants assumed upon moving. When law firm Kirkland & Ellis decided to rent 600,000 square feet in 300 North LaSalle in one of the city's largest lease deals, it doubled its net lease rate.[105] It took less space in the new building, but the firm's total rent payments were still higher after the move. Kirkland was not alone. National data reveal that over the course of the boom, office-using businesses spent an increasing share of their revenues on real estate while occupying smaller amounts of space. For example, while 1.2 percent of law firms' revenue was spent on rent in 2002, by 2009 the share had increased to 15 percent.[106] Some Chicago law firms collapsed under the weight of debt they had taken on partly to finance their relocations, casting some doubt on the "flight to quality" explanation of boom-time construction.[107]

We'll Take You Higher

Over two hundred large-scale development projects were erected in downtown Chicago between 1998 and 2009. Chicago became "a much more urbanistically appealing city" during the Millennial Boom as a result of the massive private and public investments made there.[108] Public art, free concerts, student dormitories, new retail venues and public space, and improved bike and transit access helped too.

Peeking behind the new facades, one sees a different picture. The boom was only weakly demand-driven. The city's anemic job growth and declining population should have been a drag on construction activity and property leasing. Demand for new office space emanated neither from businesses new to Chicago nor from existing business outgrowing their original space. It was those businesses and households already located in the city and region that were the primary space seekers. An executive vice president at the Alter Group, a local developer, noted aptly, "Landlords are moving around businesses that are already here . . . There's very little fresh meat."[109] Similarly, international and domestic in-migrants were not the primary source of demand for the high-rise condominiums, whose median sale price was $398,000 at the boom's peak.[110] The "new" downtown condo dwellers moved there from other parts of Chicago and its suburbs, or they were not homeowners at all, but investors hoping to cash in on the run-up in housing prices.

The amenities provided by the new buildings were appealing to occupants, but their bottom-line utility was questionable. Technological innovations in office buildings, for example, were less radical than in previous booms. The skyscrapers constructed during the Millennial Boom were taller, skinnier, and more efficient than those from previous years, but they did not revolutionize building construction in the same way as, say, the adoption of electric elevators and advanced temperature controls in the booms of the early part of the twentieth century.

What, then, prompted thousands of businesses and households to exchange older buildings for newer ones? Were buildings in the historic core suddenly afflicted with a case of mass obsolescence? The following two chapters offer a more compelling explanation of this building boom, retelling the story of downtown's expansion with a focus on the supply-side actors who facilitated the tenant relocation and commercial development processes. Responding to signals from financial markets and intermediaries such as brokers, appraisers, and the local government, developers operated according to a logic that, although not entirely removed from the interests and needs of occupants, became increasingly remote as a speculative bubble developed.

5

Who Overbuilt Chicago?

Sam Zell may be the second most famous real estate mogul in the United States after Donald Trump, even though Zell is worth more and is no less quotable. Instead of Brioni suits and comb-overs, the Chicago-based billionaire is known for his elfin stature, his penchant for motorcycles and expletives, and his iconoclastic opinions about the workings of capitalism. Like Trump, he takes every opportunity to market his name—"the most important asset I have."[1] Without a reality television show as an advertising vehicle, Zell has had to seek out other media strategies—such as buying the Tribune Corporation, which owns his hometown newspaper.

Steering clear of construction, Zell made his reputation as a "grave dancer": someone who buys distressed assets at bargain-basement prices during periods of crisis and unloads them at inflated ones once markets have recovered. He did so well as an individual investor that, in 1997, he offered shares in a new fund, called Equity Office Properties (EOP), to the public. In doing so he helped popularize the publicly traded Real Estate Investment Trust (REIT) model as a way of making commercial real estate ownership available to the average investor.

Zell went on to become the industry's chief aggregator. He borrowed heavily to purchase not just individual buildings, but the entire portfolios of his competitors.[2] At its peak EOP owned 721 office buildings in New York, Boston, and Los Angeles and their suburbs, becoming the largest REIT in the country.[3] But Chicago was always EOP's prime market. From the Chicago Mercantile Exchange to suburban Oakbrook Terrace, Zell owned almost 5 percent of the region's total office space.

By the turn of the millennium, Zell had amassed such large and diverse holdings that some observers snidely suggested that his empire had come to resemble a big, clunky "utility, with its fortunes tied to the ebbs and flows of the economy."[4] In the property world, the comparison was a stinging insult. It insinuated that Zell had lost his taste for the fast-paced, often counterintuitive transactions that characterize the most risky and lucrative segments of real estate investing. It implied that Zell had grown stodgy, an owner in it for the long term and dependent on the productive economy—as opposed to the fictive, financialized one.

Had Zell lost his edge? Was he really no longer a "vulture," or was he up to something bigger? In the role of landlord, Zell was forced to generate returns the old-fashioned way: through each building's income and operating efficiencies rather than through financial engineering. He thought that his famous brand would help him raise the rents of his highbrow tenants such as AT&T and PricewaterhouseCoopers. Zell's company offered special touches such as car washing and concierge services at his buildings. It tried to speed deals by cutting the length of its average lease in half. EOP's chief financial officer justified such perks as a way to differentiate its product: "We're trying to take it [real estate] as far away from a commodity as we can."[5] On the cost side, Zell bet that economies of scale would wring out savings. His REIT negotiated cheaper bulk supply contracts, sought lower-cost financing, and consolidated its leasing and management functions.

The response to these changes was underwhelming. Even though it outpaced the Standard & Poor's 500, EOP's performance during the early 2000s trailed that of its competitors. Despite aggressively raising rents to justify its acquisition prices, EOP could not generate enough operating profit to cover dividend payments, which it first cut and then simply ceased offering altogether. EOP then began to sell entire buildings, quietly exiting cities such as Nashville and Salt Lake City where its presence was smaller.[6]

Some observers insisted that Zell was not straying from his new identity as a caretaker, but that he was sensibly streamlining his coast-to-coast holdings so that they would be easier to manage. Others saw a return to his roots as a grave dancer: standardizing the product line would make the REIT easier to value for a takeover. They did not believe that Zell could resist the lure of the fast money. The role of a landlord lacked the adrenaline-fueled excitement and the visibility of deal making.

Zell's true path became apparent in 2006, when he announced the sale of the entire EOP portfolio to the Blackstone Group, a private equity firm based in New York. He sold his holdings for a whopping $39 billion—the largest private equity transaction in history and one that came to signal the transition from boom to bubble.[7]

In contrast to Zell's short-term experiment with long-term ownership,

Blackstone had always been known as a "chop shop" staffed by "pinstripe-clad calculators" who recapitalized underperforming properties, raised rents, and flipped them to the next group of investors.[8] Indeed, as soon as it purchased the EOP portfolio, Blackstone put it on the auction block.[9] As a result, EOP's shareholders and Blackstone made out very well, while many of the firms that purchased those assets—including Tishman Speyer and GE Real Estate in Chicago—struggled to sell as the market collapsed around them.

What happened to these buildings was not Zell's concern. He was too busy tallying up his proceeds from the deal, which amounted to more than $700 million. The largest real estate portfolio sale in Chicago history had the effect of pushing prices to heights never before seen in Chicago and in the other cities where Zell owned buildings. The years 2006 and 2007 went on record for the greatest number and volume of commercial real estate transactions in U.S. history.

Underwriting the Boom

Just as Sam Zell could not resist the urge to sell during what, in retrospect, was a speculative bubble, other building owners were unloading their holdings during what they knew was a finite window of opportunity. Potential suitors on the buy side—REITs, institutional investors, private equity funds, and high-net-worth individuals—were eyeing Chicago's building stock hungrily. They could use financial instruments such as commercial mortgage-backed securities (CMBSs) to leverage their acquisitions to the hilt. Rates of property appreciation far outpaced both the cost of capital and the profits generated by a jittery stock market. The strategy pursued not only by Zell, but also by Buck, Trump, and many other less flamboyant city builders, was clear: buy and sell.

The debt-fueled acquisitions market—not the lackluster occupant one—was the real driver of the commercial construction boom in downtown Chicago. Developers could get funding from private and public sources as easily as investors could. And, just as importantly, they knew they could sell whatever they put up. But, oddly, the growing scope and potency of global financial markets is rarely acknowledged in narratives of this era. The public just "stopped asking why. It seems really odd at first to have these empty buildings selling for record sums, but then it just became kind of normalized."[10]

In explaining why building construction continued and sale prices increased despite relatively flat growth (the signs of which were obvious as early as 2003), some pointed to Chicago's aggressive development culture. "Perhaps this oddity arises from Chicago's nature as a pro-development city in the long languishing Midwest," wrote one reporter, puzzled by this seemingly irrational behavior.[11] Indeed, finance unfolds in very localized and humanly scaled settings; every city hosts its own institutions, its particular professional networks, and its specific channels for transmitting the economic information that keeps capital mobile and makes buildings fungible.[12] While major trading centers such as London and New York City attract attention, they are merely the last link in a chain that standardizes and moves capital from smaller, diverse settings to investors in these metropoles.[13]

Some local actors felt threatened by what they saw as a new culture of high-stakes global real estate investment. Noted one banker, "These old-school, Chicago real estate guys were shut out. It [the buying and selling of office buildings] was no longer *their* game. It was a finance game, and it was played over their heads."[14] But even the old guard was instrumental in creating the conditions required for financial flows to gain a foothold in the city, facilitating deeper connections to Wall Street and the European and Asian markets. In this chapter I discuss how Chicago-based financial institutions such as LaSalle Bank generated the critical inputs, such as mortgages, necessary for yields to flow to CMBS bondholders around the globe. Brokers, lawyers, and city planners in tight-knit professional communities sanctioned the use of these instruments locally and encouraged their proliferation.

The City of Chicago also lent its legal authority and its access to cheap debt to the creation of new assets. Chicago was considered a city where property values were safeguarded by a sympathetic municipal administration. The financialized strategies adopted under Richard M. Daley were quite similar to those of developers and investors: the city took advantage of escalating property values to borrow heavily, build, and sell off assets and income streams. In doing so it facilitated global capital switching into local real estate and nurtured a deal-oriented development culture that favored the fast and the new at the expense of the old and the slow.

AN UNLIKELY BOOM TOWN?

In the beauty contest that is the pursuit of global capital, Chicago has not done as well as its peers. Foreign and domestic investors habitually leapfrog over the Midwest on their prospecting tours, settling instead on "A-list" coastal cities such as New York City, Washington, DC, Los Angeles, Boston, and San Francisco. For example, "large cap" office REITs—those that own over $5 billion of assets—maintained a small presence in Chicago over the Millennial Boom.[15] Sam Zell controlled most of this value (20 percent of EOP's holdings were in Chicago) until the sale of 2007. Chicago assets were used by investment funds as filler, to diversify and hedge against the risk of overinvestment in other, more popular locations.

Big money's avoidance of Chicago was based on a potent mixture of performance and perception. During the boom the top-tier cities consistently outperformed Chicago—Manhattan office towers, for example, sold for $280 more per square foot than those in Chicago.[16] National indices measuring the ten-year returns for different asset classes put Chicago close to the bottom.[17] Buildings in other global gateways charged higher rents and experienced faster rental growth. Gross asking rents for Class A and B office buildings in Chicago hovered around $25 per square foot—significantly lower than what landlords charged in New York City (where $100 was not unusual), Los Angeles, and Washington, DC.[18] The primary source of this sluggishness in lease rates was a lack of job growth. Adding insult to injury, commercial real estate taxes as a share of rents were some of the highest in the nation.[19]

Another reason for the city's relatively poor performance chafes uncomfortably against what many believe to be the very symbol of its success: its new buildings. Chicago historically has been known as a city that constructs too much, too quickly, a flat palette on which grandiose visions can be given material form.[20] One investor concluded that "the history of the Chicago market is an imbalance of supply and demand."[21] Another concurred: "In some ways it reminds me of Texas—it is not as wide open in zoning as Texas, but it is more wide open than most of the country. There is a natural fear that Chicago will keep building and building and get overbuilt."[22] Too much new supply threatens investment performance by reducing rents and sale prices.

More supply-constrained cities register higher rent and price increases because tenants and investors have fewer options; demand builds, pushing prices upward.[23] Chicago has a more elastic supply. It is less physically constrained than the island of Manhattan and less publicly managed than the fragile topography of San Francisco. Moreover, Chicago's lower densities provide more space for growing up and out than in coastal cities. Before the Millennial Boom, for example, Chicago's downtown was ringed by warehouses, underutilized infrastructure, older two- and three-flats, and low-rise retail buildings. These structures presented more redevelopment opportunities than in Vancouver or Manhattan, whose downtowns were forested with towers already capitalizing on their high land prices.

Prime buildable land in Chicago was available and relatively inexpensive to acquire, partly due to public policy.[24] Local governments control the pipeline of new assets and have the magical ability to make developable land appear where there was none before. Mayor Daley's primary aim in the years leading up to the boom was to open up areas on the downtown fringes for new development, unlocking the value of areas that for decades had separated the Loop from adjacent residential areas. Impediments to new development included public housing as well as rail lines, rivers, and highways. Under Daley the city demolished public housing, bought and sold land below market value, negotiated air rights with the rail companies to deck over tracks— literally creating new space out of thin air—and used property tax revenues to construct the infrastructure (bridges, transit stations, street extensions) that would rationalize and direct traffic flows to an extended "Super Loop."

The pipeline of new supply was fed by a sizable fraternity of real estate and financial professionals. Nearly 350,000 individuals worked in the commercial real estate sector in Chicago at the peak of the Millennial Boom.[25] Chicago's position as regional headquarters for real estate intermediaries and related financial services made its share of FIRE employment higher than the national average.[26] And yet, despite its size, this group of developers, financiers, and brokers constituted something of a tight-knit club.[27] With the exception of New York's Trump, Houston-based Hines, and Cleveland's Forest City, most of the developers behind the Millennial Boom had long-standing roots in the city, including local firms John Buck Company, the Alter Group, and Fifield. The most successful real estate professionals were often the progeny of other ones.[28]

The organizational culture of these Chicago actors has historically been insular and relationship-driven, which is both a cause and an effect of the city's being passed over by the largest global players. A dynamic structure of networked relationships linking developers, sources of capital, and brokered services evolved in the city over time, trading not only buildings but also personnel and political favors. In some cities these actors behave in more autonomous and impersonal ways; for example, cities such as Phoenix and Dallas are described as "transactional" cities where you "just swoop in, do your deal and get out. After you leave, no one remembers your name."[29] But in Chicago, peer groups, political affiliations, and social connectivity (all slow-filtered through race, religion, and ethnicity) influence whether your deal gets done and on what terms.[30]

These kinds of relationships are expected of smaller cities. But Chicago is not a bit player: it hosts the third largest office and residential markets in the country. Chicago's downtown market is larger than those of the eight most populous midwestern cities combined.[31] Chicago accounted for at least half of the volume of office sales in the Midwest for every year of the boom and posted consistently higher sale prices than any other city in the region.[32] Investors describe it as "institutional grade" in that the city "can attract [pension] fund money."[33] Others note that "it is one of the most liquid downtown markets in the country" and that it is a "heavily brokered city that just keeps developing and growing."[34] Part of the attraction is low barriers to entry: smaller investors can afford the purchase price of an office building in the Loop, whereas they would not be able to break into Washington, DC, or San Francisco. Capital shut out of other, more expensive markets habitually finds its way to the Windy City, usually after a short lag.

The city's sense of itself is always at risk of being pulled down by past associations (with autocracy, corruption, provincialism), but it contained enough positive symbols and promise to keep the capital flowing during the Millennial Boom.[35] Local, national, and global investors were on hand to underwrite each of the critical phases of a building's financial life cycle described in chapter 2: construction, recapitalization and acquisition, and securitization. Each of these phases reinforced the others and helped set Chicago's course toward overproduction.

CONSTRUCTION AND OPERATING FINANCE

Chicago's banks were initially inhibited from growing by New Deal-era regulations limiting their ability to expand within or across state lines. Nowhere were state-level restrictions more onerous than in Illinois: unit banking laws kept lenders operating within a single county.[36] With deposits low, retail banking growth was effectively curtailed. Even in the 1980s and 1990s, when other states relaxed these controls, Illinois bucked the trend.

Because Illinois came late to deregulation, Chicago's banking sector was made up of small and medium-sized regional depository institutions, defined as banks holding less than $100 million in assets. At the turn of the millennium, the city was considered the most fragmented of any major bank market in the country, with only 40 percent of its assets owned by the top three nationwide bank holding companies.[37] It was also spatially fragmented because of branching restrictions that kept retail outlets in the locations where they had the greatest chance of maximizing deposits. Most of the commercial banks maintained headquarters in the Loop, particularly in the historic financial corridor anchored by LaSalle Street, or were located to serve particular geographies, commercial sectors, or ethnic groups.[38]

In the late 1990s yields from commercial lending and fixed-income investments were lackluster, and low interest rates and the repeal of the Glass-Steagall Act created competition to originate loans with nonbank entities such as Wall Street investment firms. To survive, middle-market lenders relied on their retail networks for new loan originations and often entered the construction niche. Construction loans exposed lenders to risks but provided higher yields. Whereas commercial banking relationships often took years to mature into loans, real estate development projects were viewed as "one-offs" that could quickly inflate a bank's earnings figures. Loan officers were handed quarterly targets for construction loans to place. Noted one commercial mortgage broker in Chicago:

> [The banks] still have money they need to put out. So they get into construction loans. But now there's twice the money facing half the deals, so you *really* need to push it down the risk spectrum. They did spec, they did neighborhood strip malls, they did condos, they did everything. And these were smallish banks. They really didn't know how to price the risks.[39]

A second segment of Chicago's banking sector included a handful of larger regional banks that were able to expand because of the city's relative size, wealth, and penchant for construction. Continental Illinois, American National, Harris, and LaSalle became homegrown powerhouses underwriting the postwar waves of downtown building and were markers of a new era in the city's financial self-sufficiency. Like their smaller siblings, these banks excelled at the kind of "relationship banking" that cultivated tight-knit connections with clients. They too, however, often resorted to high-risk strategies to meet earnings goals. Continental Illinois, for example, aggressively expanded into oil and gas loans with funds raised by selling short-term certificates of deposit on the wholesale money markets (as opposed to relying on retail deposits).[40] After becoming the sixth largest bank in the United States, it eventually succumbed to what, in 1984, was the largest banking failure in U.S. history.

LaSalle Bank was the most conspicuous real estate lender in Chicago during the Millennial Boom. Chartered in the city in 1927 and occupying an Art Deco building at LaSalle and Adams, the bank grew so steadily that it was acquired by Netherlands-based banking giant ABN in 1979. Under its new ownership, the bank, which retained its name and identity, went on a regional acquisitions spree. In 2007 it was Chicago's largest bank, with more than one hundred branches in the Midwest, over $125 billion in assets, and fifteen thousand employees in the Chicago area.[41]

Most developers in the city had direct banking relationships with LaSalle, as commercial real estate constituted over 30 percent of the bank's loan portfolio.[42] Noted one broker, "Norm Bobbins [president of LaSalle from 1991 to 2007] was like the godfather of the Chicago real estate. Everybody knew him, and everyone had a relationship with his bank."[43] The company was purchased in 2007 by Bank of America, which publicly voiced discontent that LaSalle had underwritten so many bad construction loans.[44]

Local bank Corus started off as a student loan and mortgage firm, but shifted into commercial real estate lending during the Millennial Boom, making construction loans first on hotels and office buildings and later on condos. In 2009, construction loans accounted for 88 percent of Corus's outstanding loans, the largest share for any U.S. bank with more than $100 million in loans.[45] After lending money to Chicago developers, the bank looked south to the burgeoning condo markets of Miami and Atlanta. Investors questioned that strategy, betting that the condo bubble would burst by shorting Corus's stock. The bank eventually failed in 2009 and became the poster child of an overextended heat-seeking industry.

LaSalle and Corus were highly exposed to construction and land development projects, but they were not alone. In 2005, 36 percent of all Chicago bank loans were in commercial construction loans, commercial mortgages, and apartment building mortgages (table 5.1). This figure represented an increase of ten percentage points from the downturn of 2001—compared with a three percentage point increase for the nation as a whole.

TABLE 5.1. Chicago's Largest Banks and Their Concentration in Real Estate Loans, 2006 and 2007

		2006			2007		
	Size ranking	Assets (billion $)	Real estate loans (billion $)	Share of assets (%)	Assets (billion $)	Real estate loans (billion $)	Share of assets (%)
LaSalle Bank National Association	1	73.0	20.2	27.6	74.4	23.4	31.4
The Northern Trust Company	2	52.3	3.8	7.3	58.4	4.0	6.8
BMO Harris Bank National Association	3	41.8	13.9	33.2	41.5	15.5	37.4
Corus Bank, National Association	4	9.8	4.1	41.4	8.8	4.3	49.5
MB Financial Bank, National Association	5	7.6	1.4	18.2	7.8	1.0	12.6
Park National Bank	6	3.7	2.8	76.3	4.2	3.1	72.4
Cole Taylor Bank	7	3.4	1.8	53.5	3.5	1.8	49.5
The PrivateBank and Trust Company	8	2.9	1.8	62.7	3.4	2.1	63.1
ShoreBank	9	2.0	1.0	47.8	2.2	1.1	50.1
Marquette Bank	10	1.6	0.9	53.3	1.6	0.9	57.0

Sources: Data from Federal Deposit Insurance Corporation (FDIC), Statistics on Depository Institutions (SDI) and Uniform Bank Performance Reports. For 2005, see Steve Daniels, "Only game in town: Banks bet big on real estate" *Crain's Chicago Business*, July 23, 2005.

A third segment, money center banks, had its own interests in moving into Chicago real estate. Although Citigroup, JPMorgan Chase, and Bank of America had more balanced portfolios than the smaller banks, they entered into loan origination when their fee-based securities work became more lucrative.[46] They bought out smaller commercial banks to acquire origination capacity and to prime the supply of mortgages that could be placed with their securities divisions. Chicago lenders were vulnerable to takeovers because they were in a less concentrated market yet individually well capitalized; they were, without exception, on the losing side of the merger and acquisition wave that began in the 1980s.[47] While coastal cities and historic banking hubs such as Charlotte, North Carolina, won out, the consolidation left Chicago as one of the few major U.S. cities that lacked even one headquarters of a money center bank.

This consolidation did not, however, affect the ease with which Chicago-based investors and developers could borrow. Money center banks provided syndicated financing to the most trustworthy producers of trophy buildings. To build up Wacker Drive, for example, major developers could "call up New York and get whatever money they wanted. The Bucks and Hineses of the world build what and when they want."[48] In addition, major REITs such as Vornado, GE Capital, and BlackRock, as well as pension funds such as the California State Teachers' Retirement System (the second largest public pension fund in the United States) made construction loans for the new West Loop towers. Enticed by fees and seeking to pump up their valuations, foreign financiers entered the market. During the Millennial Boom most non-U.S. investment in Chicago came from continental Europe, the Middle East, and Canada (a switch from the previous construction boom of the 1980s, which was fueled by Japanese and British capital).[49]

The power to intensify or stave off overbuilding lay in the hands of these three kinds of financial institutions. Developers assumed they would turn off the spigot as soon as signs of saturation were spotted. Indeed, the dominant tropes of the late 1990s buildup were that banks had learned from the mistakes of the 1980s and that contemporary demand for construction financing was empirically different from that in the previous boom. As building heated up in 2000, banks raised interest rates and tightened underwriting standards. The managing director of Midwest real estate lending for Bank of America boasted that he had denied financing to two significant condo projects in Chicago: "The condo side is troublesome. We're being much more cautious about the level of speculation."[50] Banks also tightened credit to office developers, requiring lease commitments for at least 30 percent of a proposed building.[51] "Today, developers must have at least one anchor tenant to secure financing . . . The era of the see-through building is over," declared another lender.[52] The message to developers was, quoting a newspaper article from that time: *Cool your jets.*[53]

Concerns about credit quality, however, were quickly shelved to make way for the wave of global capital looking for quick placements after the recession. Banks loosened their requirements, assisted by other real estate service providers who loosened theirs. Appraisers, for example, were in such high demand that they yielded to client pressures for higher valuations. If they did not succumb, they feared losing business to the many new entrants in this increasingly competitive field.[54]

Speculative buildings—those with no or few tenants or buyers committed before the developer breaks ground—found financial backing. Perhaps the most scrutinized case of a speculative building was John Buck's addition of another trophy, 155 North Wacker, to his collection in the West Loop, as described in Chapter 4. When Buck received construction funding from LaSalle Bank, Bank of America, and German financier Helaba, and it was clear that banks were willing to take such risks, other developers jumped into the fray. At the height of the boom, pre-leasing requirements became more of an irritating formality than a do-or-die decree. "If you don't have 50 percent pre-leasing, it just means you have to put in more equity," noted one financier.[55] So strong was their faith in their product's ability to create its own market that developers moved into even riskier territory, like taking on expensive mezzanine financing. One developer noted that he "would have jumped out of an airplane without a parachute" before 2008.[56]

And when they did not attract a sufficient number of office tenants or condo buyers, developers played games to work around underwriting standards. They sold condos to themselves and to brokers, for example, or offered off-the-books discounts to buyers who signed purchase contracts before construction began (i.e., "early bird financing").[57] Anyone with earnest money was fair game as a pre-sale, and lenders often closed their eyes to inconsistencies or less-than-needed incomes. Under pressure to pump up quarterly earnings, banks progressively loosened their requirements to move their money out the door. They forecast their earnings based on recent instead of future rental income, ignored free rent and concessions in the valuations they commissioned, and sped up closings.

These tactics were mutually beneficial for banks and developers. Most condo developers sold enough units to pay off their construction loans at maturity, usually two to three years after breaking ground, when the majority of buyers had closed on their units. In the case of several spec office buildings, the gamble paid off; post-construction, Buck and Fifield were able to attract anchor tenants such as Humana and Ernst & Young. Even in the worst cases, banks reasoned, they could always sell these properties to someone else.

ACQUISITION

Chicago developers and construction lenders leaned heavily on two "insurance policies" as they sunk more caissons into the ground: secondary markets for the buildings and for the debt secured by those buildings. The willingness of investors to take new assets or liabilities off their hands encouraged lenders to continue to originate loans and developers to build long after occupant demand for the assets had dried up. The trick was to sell at the right time.

Office workers in the Loop went about their business unaware of the Monopoly game being played with their cubicles. Almost every office building in the Loop—85 percent of the stock, or 83 million square feet—sold at least once during the Millennial Boom.[58] Sales included the new towers along Wacker Drive as well as older landmarks such as the Sears (now Willis) Tower, the Wrigley Building, and the Civic Opera Building.

The pace of transactions was frenzied, peaking in 2006. That year also represents the high-water mark for volume: office sales reached a record $5.5 billion (fig. 5.1). In 2007 Sam Zell completed the blockbuster sale of EOP to

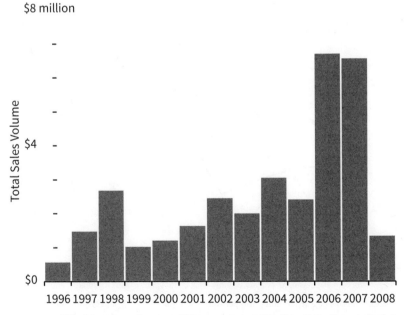

FIGURE 5.1. Office sales volume, downtown Chicago, 1996–2008. Data from CoStar Property Analytics.

Blackstone; that year's total, at $5.1 billion, also encompasses the almost immediate resale of the EOP portfolio to Tishman Speyer—a firm that thereby became one of the largest commercial landlords in Chicago overnight.[59]

Local REITs such as the Prime Group and Inland were active buyers, but new entrants such as Parkway (from Jackson, Mississippi) and Piedmont (from Atlanta) moved in during the peak years. In the East Loop even "small cap" REITs, those valued at or below $1 billion, could pick up older Class A and B buildings (fig. 5.2). Private equity funds such as Blackstone vied for the same pool of buildings as the larger REITs. But these funds had shorter time horizons: they were looking to hold onto the buildings for about three years instead of the ten-year holds of most trusts. Some were started by experienced real estate firms such as Buck, Hines, and Tishman Speyer, while others were straight money managers for high-net-worth individuals and firms with cash on hand.[60]

All of these players, but especially the out-of-town money managers, relied heavily on local brokers and appraisers to advise them as to which buildings they should purchase or sell at what prices. These intermediaries provided the on-the-ground information about tenant lease agreements and ownership structures that was necessary to trade assets.[61] For example, Holliday Fenoglio Fowler handled the disposition of assets from Blackstone

FIGURE 5.2. Average acquisition prices per square foot for Class A, B, and C office buildings, downtown Chicago, 2000–2010. Data from CoStar Property Analytics.

following the EOP sale. They not only prepared valuations and marketing plans for the twenty-three buildings Blackstone purchased in Chicago, but also made sure that the portfolio was legible to distant investors: "We had to make sure everybody agreed on value."[62]

Investors were initially drawn to those new Class A office buildings with creditworthy tenants on long-term leases. Most of the buildings located in the West Loop were picked off quickly by the largest funds and trusts. Some trophy buildings sold several times during the boom. Because of the competition for assets, purchase prices for Chicago office buildings increased steadily between 2001 and 2008. Prices oscillated around $325 per square foot for Class A office space during this time, but some of the highest individual prices ever recorded in the city were recorded during the bubble years of 2006 and 2007 (see fig. 5.2).[63] Noted one real estate analyst:

> There's so much money from capital markets looking for investments here that actual lease rates have become less important. Now there's money from outside [the city, country] investors. They're looking at location and future cap rates more than leasing ... Look, 190 South LaSalle had hardly any tenants when it sold but it still sold for so much money.[64]

Even buildings left over from previous booms—such as the John Hancock Center and One Prudential Plaza—sold at what observers called "premium prices" ($383 million and $470 million, respectively) despite major tenant losses.[65] Investors were aware of the rising vacancies but continued to bid up prices in spite of them. A speculative bubble buoyed by hot air, after all, was still better than depreciation: "Even for a deal that's not stable, it allows you to push pricing that much more."[66] But others tentatively voiced concerns. One investor who bypassed Chicago to purchase an office building in Dallas noted, "I'm amazed at the prices being paid compared to the real estate fundamentals [here]. In 2006 a wave of capital has focused on global U.S. cities, and Chicago is certainly one of them."[67]

Capitalization rates—the ratio of a building's rental income to its sale price—continued to plummet, reflecting the fact that prices were moving out ahead of rents. Average cap rates started off at about 9 percent at the beginning of the boom and sank below 5 percent in 2006. In their analysis of the Chicago market, Dermisi and McDonald found that occupancy was a statistically insignificant factor in determining the sale prices of Class A buildings.[68]

Like Sam Zell, most landlords knew that the "best way to unlock [the] value was to transfer the assets to other players."[69] They employed gimmicks to make their buildings look better on paper. Noted one investor:

When rents are low or tanking, everyone becomes a buy-and-flip outfit—
even folks who have historically been buy-and-hold 'ems. The buy-and-flips
offer all kinds of tenant improvements and concessions so they can keep tenants long enough to make the next sale. The buy-and-holds . . . well, they just
have to eat those losses.[70]

Owners "salted" their leases and sales by offering exceedingly favorable
terms, or they misrepresented the amount of space leased or rental income
collected.[71] Appraisers raised their valuations upward under pressure from
their client banks, which wanted to move larger amounts of debt out the door
and meet their own quarterly goals. As one appraiser noted:

No one had the time to read the final report anyway . . . Everything was just a
push beyond that "range of reasonableness." You felt it on the phone, talking
to people. You were always looking for that next cap rate or comparable that
would allow you to make it to that high target. If you couldn't get there, the
borrower couldn't get the loan, the buyer couldn't execute the purchase, and
the bank would just go with someone else the next time.[72]

Sluggish rental growth and truncated tenant rosters should have diminished investors' interest, but demand for Chicago office buildings appeared
insatiable. Capital supply drove both construction and investor demand for
the new stock. During the Millennial Boom the acquisitions market was flush
with cash: REITs sold shares through stock offerings, and pension funds
threw money at private equity funds. But mostly what these investors did was
borrow. REITs leveraged their acquisitions; the country's fifty largest trusts
carried approximately $270 billion in debt on their balance sheets in 2008
for a total leverage ratio of 70 percent.[73] Banks' enthusiasm for real estate
was stoked by the fact that they could repackage these loans as securities and
earn fees selling them to investors in secondary and tertiary markets far from
Chicago.

SECURITIZATION

Chicago was no stranger to securitization. Agricultural commodities were
first traded in the city's famous exchanges in the mid-nineteenth century.
To smooth out seasonal fluctuations in prices, futures and options contracts
on these products soon emerged.[74] The leap into financial commodities occurred only in the 1970s, when contracts similar to those used for wheat and
corn allowed the trading of foreign currencies at the Chicago Mercantile Exchange. After that the Chicago Board of Trade began selling options contracts
for stocks and single equities and subsequently for futures on interest rates

and U.S. Treasury bills. The year 1982 saw the value of trades in financial futures exceed those derived from agricultural products.[75]

The securitization of real estate also got its start in the city. Mortgage bonds were first issued in 1893 by the local Peabody Houghteling Company.[76] In the 1920s Chicago's First National Bank became the first bank to organize a "security affiliate," the First Trust and Savings Bank, to sell and manage bonds secured by its real estate loans to raise more capital for the bank.[77] This was considered a revolutionary shift in banking, and the "Chicago Plan" was soon copied by New York–based financial institutions and others. The ability of banks to sell both mortgages and the securities built on them was one of the reasons for the credit expansion of the Roaring Twenties, the banking collapse of the 1930s, and the subsequent passage of New Deal legislation segregating commercial from investment banking. In 1988, more than ten years *before* the repeal of the Glass-Steagall Act, Continental Illinois Bank issued securities representing interests in a pool of leveraged buyout loans.[78] This led some to credit the bank with issuing the first collateralized debt obligation (CDO).[79]

Despite its Chicago origins, however, the securitization field's center of gravity moved east, and by the 1980s it was firmly rooted in Wall Street.[80] Little of the lucrative fee business was transacted in Chicago, although several New York firms did open Chicago offices to source the loans they would ultimately securitize. Noted one Chicago banker, "We didn't want to do anything we didn't understand. The Chicago banks mostly stayed on the sidelines during the derivatives rush."[81]

The sole exception was the same regional bank that had provided much of the construction financing to Chicago developers over the Millennial Boom: LaSalle. As table 5.2 shows, LaSalle housed one of North America's largest CMBS commercial conduit groups, completing more than $8 billion in loan origination and securitization in 2006 alone. With its CMBS division employing more than seventy individuals at its peak, LaSalle stood out as the only hometown bank that was "cool enough to play with the New York boys."[82] Few of the other Chicago banks became large conduit lenders because of the associated costs and risks, although some regional banks, such as Cleveland-based Key Bank and Pittsburgh-based PNC, dabbled.

Like its middle-market peers, LaSalle invested heavily in its relationships with local developers and investors (such as General Growth Properties REIT) and, in the words of a competitor, "would have done anything for its clients even if it involved throwing risk out of the window."[83] CMBSs allowed LaSalle access to larger pools of capital at better rates. Chasing fee business and selling off pieces of its exposure allowed the bank to profit from an ever-

TABLE 5.2. Top Loan Contributors to CMBS Deals, United States, 2005–2007

	2005		2006		2007	
	Total (million $)	Share of total (%)	Total (million $)	Share of total (%)	Total (million $)	Share of total (%)
Wachovia	16,201	9.9	21,890	10.9	24,175	10.8
Credit Suisse	12,751	7.8	16,009	8	14,730	6.6
Banc of America	13,921	8.5	13,871	6.9	15,624	7
JPMorgan Chase	9,933	6.1	12,495	6.2	11,948	5.3
Lehman Brothers	8,617	5.3	11,666	5.8	14,794	6.6
Deutsche Bank	9,956	6.1	11,269	5.6	9,593	4.3
Morgan Stanley	9,689	5.9	10,665	5.3	13,785	6.2
Merrill Lynch	6,768	4.1	8,560	4.3	8,642	3.9
LaSalle Bank	4,366	2.7	8,264	4.1	6,811	3
RBS Greenwich	7,726	4.7	7,882	3.9	9,175	4.1
Bear Stearns	6,886	4.2	7,813	3.9	9,563	4.3
Citigroup	3,562	2.2	7,147	3.6	6,917	3.1
Goldman Sachs	6,743	4.1	5,265	2.6	8,470	3.8
UBS	3,428	2.1	5,248	2.6	9,699	4.3
Capmark	2,774	1.7	4,615	2.3	2,620	1.2

Source: Data from Commercial Mortgage Alert "Top Loan Contributors," https://www.cmalert.com/rankings.pl?Q=76.

Note: Based on amount of collateral supplied to U.S. securitizations backed by recently originated mortgages. Countrywide replaced Capmark in the top 15 in 2007.

expanding portfolio of risky loans. Noted one former employee, "If the answer to the question, 'can we sell it?' was 'yes,' well, then we made the loan."[84]

Although New York City remained the headquarters for bundling and selling securities, Chicago was a robust market for originating CMBS loans. Approximately 160 CMBS loans were made to purchase or construct office buildings in the city between 1998 and 2013 (fig. 5.3). The largest borrowers were unlisted REITs and the so-called bulk buyers of portfolios, such as private equity funds.

Borrowers were drawn to the favorable terms and the speed of turn-around of CMBS debt financing. A CMBS loan proffered by an investment firm provided the borrower with longer-term (ten years), fixed-rate debt at interest rates typically a quarter of a point to 1 percent lower than on conventional mortgages offered by banks—all in deals that closed in fewer than twenty-five days, as opposed to the usual months. But it was the size of CMBS loans that really drew in borrowers. Almost overnight Chicago became a city where investors could purchase multi-property portfolios, whereas most previous sales had been one-offs. Closely watched peer behavior normalized the use of CMBSs; when one reputable investor used securi-

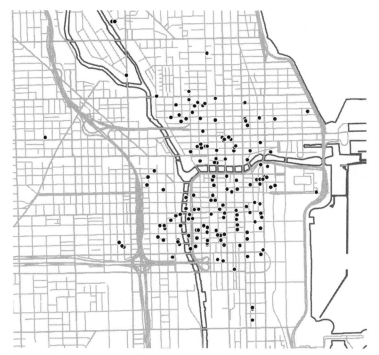

FIGURE 5.3. CMBS-financed buildings, downtown Chicago, 1998–2013. Data compiled from *Crain's Chicago Business* news reports; Cook County Recorder of Deeds, http://cookrecorder.com/search-our -records/; Trepp Analytics, Trepp CRE Data Feed; https://www.trepp.com/cmbs/data-feed/.

tized debt to purchase an office tower at close to a 100 percent loan-to-value ratio, others tried to follow suit.

Some of Chicago's most iconic office, hotel, and retail buildings were pur-chased or refinanced with CMBS loans, including the Sears (Willis) Tower, the John Hancock Center, and the Hotel Burnham. Investors purchased lesser-known Class A office towers in the West Loop and River North sub-markets using CMBS loans for up to 90 percent of the acquisition price.[85] In the end, the individual buildings were practically irrelevant to the distant packagers and purchasers of these securities. Noted a local broker:

> It's immaterial [which buildings the loans were secured by]! The Rookery [Burnham and Root's landmark on LaSalle Street] could have been made out of Legos for all they cared. If there was even the smallest amount of cash flow, it was interesting.[86]

What did matter was the specialized knowledge that local lawyers, bank-ers, brokers, and appraisers had built up over Chicago's many decades as a fi-nancial center. Although some commentators believed that CMBSs "wreaked havoc by turning a relationship city into a purely transactional one," others recognized the important role played by local firms in facilitating those transactions.[87] The institutional linkages necessary to securitize debt were opaque, so the expertise and interpersonal relationships of professionals in Chicago made clients feel secure borrowing through these less tested channels.[88] Investment banks worked closely with Chicago appraisers familiar with CMBSs and estimated that about three-quarters of their mortgages came through brokers.[89]

Some of the local knowledge provided by Chicago professionals got lost in the securitization process. To satisfy money managers, banks, insurance companies, and pension funds clamoring to purchase the bonds, the packagers removed underlying properties from their contexts and bundled their loans with other loans from different building types and locations according to proprietary formulae. Noted one banker, "These CMBS shops are staffed by 28-year-olds on Wall Street. They don't know Chicago, and they've never seen a recession . . . It's just: can they place it?"[90]

As the best buildings were sold and resold, CMBS originators cast a wider net to garner the volumes they needed to meet the demands of issuers. Conduit banks became the lenders of last resort, making funds available for questionable projects underwritten on shaky projections of future income. With over $200 billion in new issuances in 2007 nationally, hardly anyone looking to invest was turned away. "This market was on fire," commented one participant.[91] It is no wonder that loans made during the bubble of 2007 were subsequently viewed as the "worst-performing issues of all time."[92]

Planning for Expansion

The previous sections of this chapter have demonstrated how local real estate professionals establish the conditions under which property finance flows into cities. Conduit lenders like LaSalle Bank, middle-market REITs, accommodating appraisers, and hometown developers helped deepen relations between Chicago and capital markets. Their involvement helped push what started as a boom grounded in modest "replacement" demand from existing tenants into a speculative bubble in which pricing and supply had only tenuous connections to the occupant market.

Perhaps the most noteworthy agent of property financialization during this period was not a private actor at all, but rather the local government. Mayor Richard M. Daley's administration played an activist role by subsidizing an ever-flowing stream of new development. The city laid the groundwork for the Millennial Boom using policy instruments and legal powers that only the public sector could deploy: taxation and spending on infrastructure, land use regulation, and subsidies to developers and tenants. In doing so it helped create the assets that could be transacted and monetized by the actors described above. Specifically, the boom was aided and extended by two main strategies: financing mechanisms that allowed the city, developers, and corporate tenants to leverage the property base and tap the *public* debt markets to undertake ambitious building projects; and zoning modifications that maximized rentable building area.

THE POLITICS OF DOWNTOWN DEVELOPMENT

In Chicago, as in other cities across the country, downtown development is of utmost importance to the municipal government and other land-based elites—business associations, cultural institutions—that comprise the "growth machine."[93] The combined effect of powerful institutions, differentiated land uses, density, and impressive architecture creates the spectacle and the intensity that defines the city's identity. That identity, however, is always threatened by competition with the suburbs and other central cities. Not surprisingly, the downtown has become the receptacle for the municipal administration's dreams and anxieties, making it the most managed space in Chicago.

The City of Chicago assigned an elite corps of experienced administrators to oversee its planning and development operations for the estimated six hundred projects that were built in the Central Area between 1998 and 2008. They shared responsibility with the five elected representatives, or "aldermen," whose wards covered downtown.[94] Planners and aldermen also worked closely with nonprofit membership associations and interest groups, whose leaders sat on the Zoning Board of Appeals and Plan Commission.[95]

While all these actors wielded considerable influence over downtown development, there was no more powerful actor than the mayor of the city. Mayor Richard M. Daley presided over the downtown's transition from the end of the 1980s expansion (he was elected in 1989) through the entirety of the Millennial Boom. Its transformation is often attributed to him alone.[96] Stories of the mayor's omniscience about the smallest corner of the city and his strong-arm tactics (e.g., orchestrating a midnight raid to reclaim commuter airport Meigs Field as a park) contribute to this mythology. A former city planner in the Central Area remarked, "There's really only one planner in Chicago, and that's the Mayor. The rest of us are just implementers."[97]

Although Daley sought to actively manage the city's aesthetic and habitually promoted planning commissioners into his inner sanctum, he believed private real estate actors to be the only ones with the capital and technical expertise to implement his vision.[98] Only they could turn the downtown into a global center with new residential and entertainment opportunities; a college campus; a tourist and retail destination; and a symbol of Chicago's dominion over the vast stretch of land between the coasts. His aspiration was also financially motivated. Downtown was the most fiscally rich part of the city, and real estate transfer fees and appreciation would allow the municipality to sustain low tax rates, attain favorable bond ratings, and increase borrowing for expensive infrastructure schemes. The attraction was mutual: real estate professionals sought to influence the chief planner through lobbying, campaign contributions for his reelection, and gifts to his favorite charitable organizations.

Daley set his sights on the downtown's perimeter. Even during the last major growth spurt in the 1980s, the city had had little success in pushing the development frontier south and west of the historic core. Dearborn Park and Presidential Towers, two subsidized residential developments, spawned isolated residential conversions, but developers had avoided most of the Near South and Near West sides, which were cut up by railroad tracks, terminals, rights-of-way, and train servicing depots. Many of the larger switching yards and tracks were still owned by the railroads, which largely dismissed the city's repeated requests to consolidate facilities and open the land for development. This infrastructure began to lose its utility in the 1970s, leaving behind "a Chinese Wall [that] prevented the logical expansion of downtown in that direction."[99] Socioeconomic impediments had similar effects: fear of Chicago's high-rise public housing limited private residential development to discrete enclaves. How could the city's mighty economic center end in postwar housing projects and "a sorry string of sleazy bars, vacant storefronts, and 600 acres of unused rail yards?"[100]

Throughout the latter half of the twentieth century, the city was motivated by its desire to push development beyond these barriers.[101] To create more developable land, planners set about removing them.

FINANCING IMPROVEMENTS

During the mayor's father's reign, the city had been able to turn to the federal government and the civic-minded business elite to finance clearance and redevelopment. But in the 1990s, Richard M. Daley found the traditional growth coalition in disarray and Washington less willing to underwrite the fortified islands of residential development and convention centers that had characterized earlier downtown renewal. Congressman Dan Rostenkowski, who from his perch on the House Ways and Means Committee would "feed the steel and concrete beast with federal largess," was indicted for mail fraud and stepped down from public office in 1994.[102] With the exception of declining block grants and federal HOPE VI funding (which underwrote the most ambitious public housing redevelopment initiative in the United States, Chicago's "Plan for Transformation"), the city was on its own.

In response, Mayor Daley and his consultants gravitated toward an obscure economic development strategy called tax increment financing (TIF) that the state of Illinois had authorized for municipal use in 1977. TIF is a local financing tool that allows municipalities to designate a "blighted" district that would not be redeveloped "but for" public assistance. They can then use the expected increase in property taxes generated by the properties in the district to finance initial and ongoing redevelopment expenditures there. TIF became Daley's urban redevelopment strategy in the way that federal urban renewal funding was his father's favored one. Because TIF requires no approvals from higher levels of government and is drawn from the city's own revenues instead of transfer payments, the mayor had autonomy over how he used these funds, and developers had great influence over budget decisions worth millions to them in subsidies. At the urging of downtown developers, the City of Chicago designated TIF districts in places where they both stood to profit from increased land values.

Although findings of blight were a stretch in all but the earliest TIF designations, development consultants dutifully collected the data that would allow different parts of the downtown to be "activated" by TIF designation (fig. 5.4). In 1984 Mayor Harold Washington designated the first TIF district, the Central Loop TIF, so that the city could jump-start a project that, ironically, remained a problem thirty years later (Block 37, a property just east of City Hall that sat vacant for decades). The city began creating TIF districts in the South Loop in 1989, the year that Richard M. Daley was elected mayor.[103] The City Council deferred to the mayor's wishes and approved an expansion of the Central Loop TIF district to cover the area bordered by Michigan, Dearborn, Congress, and the Chicago River—the highest-valued property in the city at the time—in 1996. His administration continued to propose these overlays downtown through 2000, including the Jefferson/Roosevelt and the Ohio-Wabash TIF districts—the latter being a small district initiated by developer Steve Fifield for the sole purpose of converting the Shriner's Medinah Temple into a Bloomingdale's home furnishings store.

Whether it was intentional or simply fortunate, six of the eight TIF districts in the Central Area were in place before one of the greatest run-ups in property values ever to occur in Chicago. The funds available for discretion-

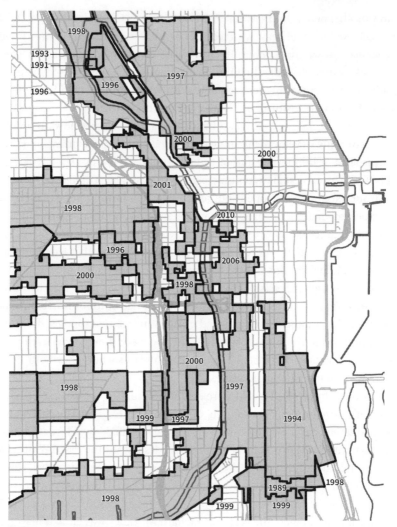

FIGURE 5.4. Central Area TIF districts and dates of designation. Data from City of Chicago Department of Housing and Economic Development.

ary public spending ballooned, enabling the Daley administration to become an important partner in the downtown's redevelopment. TIF districts generated over $600 million a year after the triennial reassessment of 2006. This amount was more than the city spent on capital improvements that year, more than the average budget of the entire Department of Streets and Sanitation, and more than twice as much as the city collected from its real estate transfer taxes in that year of peak sales.[104] With these monies, the city encouraged the downtown expansion by absorbing many of the risks of development, particularly those involved in siting commercial buildings beyond the historic core.

Approximately eighty downtown development projects received TIF allocations during the Millennial Boom (fig. 5.5).[105] In the city's eyes, "all projects have gap financing needs, and TIF completes the deal."[106] All proposed downtown developments, therefore, were automatically candidates for public assistance—no matter how profitable they were. TIF funds covered developers' financing expenses and the costs of acquiring and assembling privately owned land. They were also used to pay for infrastructure, from the most routine sidewalk repairs to major bridge and street reconstruction.[107]

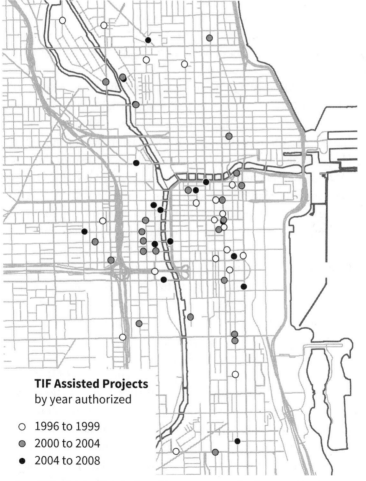

TIF Assisted Projects
by year authorized

○ 1996 to 1999
◉ 2000 to 2004
● 2004 to 2008

FIGURE 5.5. TIF-assisted projects in Central Area, 1996–2008. Data from City of Chicago Redevelopment Agreements and Annual TIF Reports.

The Near South TIF district (designated in 1990; expanded in 1994) illustrates how the city laid the foundation for development in the years before the
boom. Developers of the 80-acre Central Station project demanded approval
of this TIF district before moving forward. They had purchased land from
the Illinois Central Railroad in 1989, and TIF funds were used to deck over
the railroad tracks, which allowed the developers to exploit the air rights and
erect structures on pedestals close to Lake Michigan. Despite the approval of
the TIF district and the resolution of knotty zoning negotiations, the site lay
undeveloped from 1990 until 1996. As the boom geared up, the developers
accessed additional financing and eventually built what some have called the
largest public-private mixed-use project in the city's history: over six thousand townhouses, condominiums, and rental units in high-rise and mid-rise
buildings and 19 million square feet of offices, residences, and stores.[108]

Central Station catalyzed projects throughout the South Loop. During the
1990s the Near South Side's population grew 39 percent, to about ten thousand residents, primarily due to the development of this one development.[109]
The city spent additional TIF dollars making roadway improvements to allow for better traffic circulation around a new retail corridor on Roosevelt
Road, and developers continued to add new towers and mid-rises nearby.

While the city used TIF to help the South Loop become a premier residential location, it paved the way for the West Loop to become the new office
district. Noted one planner, "Before the last boom, City Hall knew which
areas were going to take off [for new office construction], and we focused
city services there."[110] Infrastructure improvements, such as commuter rail
station upgrades and the reconstruction of the two-story commercial access
road below Wacker Drive, encouraged the westward shift. Four West Loop
TIF districts were in place before property values spiked, producing a surfeit
of funds to subsidize new office construction.[111] TIF funds were spent on
new parks, station upgrades, and grocery stores that buffered a growing residential area in the Near West Side from the Loop. Corporate headquarters
relocating to buildings in this area also received TIF assistance.

Although he focused on expanding the boundaries of the downtown
outward, Daley did not neglect the core. The Central Loop TIF district was
the oldest and most fiscally productive of them all, and it allowed the city
to boost its appeal to prospective downtown residents through investments
in retail and entertainment projects on the Loop's eastern fringe. The city
preserved the historic facades of select vintage symbols (mainly hotels and
theaters), reignited the troubled Block 37 project, and de-malled State Street,
the modernist pedestrian and transit corridor that had failed to stem the
hemorrhaging of retailers. The city paid to preserve Louis Sullivan's famous

terra-cotta and cast-iron masterpiece that once housed the Carson Pirie
Scott department store on State Street (retailer Target subsequently moved
into the space). Even though it was not formally inside a TIF district, Mayor
Daley found ways to use $95 million in TIF funds to underwrite Millennium
Park.[112] A whirlwind of TIF activity coincided with the bubble years as the
city tried to get money out the door before the Central Loop TIF expired in
2008—coincidentally, the year before commercial construction starts plummeted by 70 percent.

Implementing the city's goal of a revitalized downtown required the complicity not only of local property developers, but also of public debt markets.
In order to fund subsidies and infrastructure from TIF, municipalities generally sell off rights to the expected streams of future property tax revenue as
debt instruments. In many ways, TIF allows local governments to mimic secondary market transformers such as investment banks; but instead of bundling and selling tranches of loans as securities, they trade slices of their own
property tax bases for up-front development capital. In another parallel with
private mortgage securitization, cities often sell these revenue streams without the knowledge of the individual property owners paying their tax bills.

In most cases municipalities sell bonds secured by a dedicated stream of
property taxes generated by and around the new development. The City of
Chicago, however, piloted a method of funding development projects in TIF
districts that distinguished it from other cities and allowed it to borrow heavily during the Millennial Boom without tarnishing its bond ratings.[113] Starting in 1996 the city began using developer notes (also called "tax anticipation
notes"), risky, unrated debt instruments that were provided to developers
on shorter terms.[114] Developers could choose to hold onto the notes and be
repaid over time with interest from the city, obtain a loan against the notes
from a bank, or "monetize" the notes, selling them in larger denominations
to investors. Proceeds from the loan or sale provided cash for the development costs.

By the 2000s global capital surpluses and low interest rates—the same
factors that made mortgage-backed derivatives so popular—created a rush
to buy these quirky public debt instruments. TIF debt was also a way for
national investors and banks to rebalance their portfolios and for local banks
to capture more City of Chicago business in the deeply entrenched networks
of rent-seeking known as Chicago development politics.[115] Whatever their
motivation, bond and note buyers saw little downside risk as property values
continued their steady ascent and the city's bond ratings improved.

Lending its good ratings to its TIF debt allowed the City of Chicago to
improve the appearance of what were essentially speculative instruments. Re-

ciprocally, the growing interest of banks, pension funds, and life insurance companies in these instruments inflated their values and gave the city the confidence to float more TIF debt. It borrowed extensively, not just through TIF notes, but also through more conventional debt instruments such as general obligation and revenue bonds. The city's bonded debt load grew by 122 percent, from $3.1 billion in 2000 to $6.9 billion in 2009, due partly to its growth-oriented practices but also to its unfunded pension liabilities and low property tax rates.[116] By 2011 just servicing its debt ate up 21 percent of the city's annual budget.[117]

ZONING FOR DOLLARS

Plans provide an ersatz everything-in-its-place kind of order. While they are not blueprints, plans signal to developers a willingness on the part of the city to accommodate specific kinds of development in particular locations.

Although he liked planners, Mayor Daley abhorred plans.[118] Instead, financing drove development priorities as the city offered TIF assistance to projects initiated by developers. The most recent master plan for the Central Area had been released in 1983, in preparation for negotiations over the World's Fair. Daley allowed his planning department to map out "expansion zones" to push the boundaries of the Loop outward and to propose marginal improvements on the fringes, but in general, the mayor avoided any written proclamation of a future vision that could potentially tie his hands when responding to developer requests.[119]

After a three-year planning process, Daley grudgingly released the 2003 *Central Area Plan*. The plan was primarily a response to the conversion of Class C office space to residential uses. Printing factories in the South Loop had been converted into apartments in the 1980s to widespread acclaim, but those buildings were outside the boundaries of what had traditionally been considered the Loop. Allowing residential conversions in the inner core elicited strong disapproval from commercial property owners and developers, who feared they would soon be zoned out. Moreover, the West Loop, the logical location for new office construction, was originally zoned for lower densities that would restrict developers' profit calculus. Developers put pressure on the city to produce a plan that would balance competing land use claims.

The 2003 plan provided a broad framework for new transit stations, green infrastructure, and open space projects. It emphasized the value of mixing land uses, but also of concentrating office towers in the core and on the west bank of the Chicago River. It showed high-density mixed and residential uses lining the eastern and southern perimeters of the downtown. While this plan provided a framework for future development, it was, like most plans, aspirational; it lacked the power to compel private development to occur in particular locations or to prevent unwanted development in others. Many of its recommendations were ignored during the climax of the building boom, which occurred while the growth coalition was distracted by plans for the 2016 Olympics bid. Daley refused to embrace the plan, and it collected dust until 2008, when city administrators hired consultants to draft the *Central Area Action Plan* (CAAP), which was a blueprint for implementing the 2003 plan.

The CAAP and the 2003 plan are aspirational in another sense of the word: they are unified in expressing the belief that downtown growth is not only possible, but limitless. The 2003 plan anticipates that 1.6 million square feet of new office space could be absorbed every year for the following twenty years. Although the CAAP notes in passing that downtown employment has declined every year since 2000, it nevertheless repeats a commonly heard estimate of "6,000 new jobs per year in the Loop" and projects deficits in office space amounting to almost 25 million square feet by 2020. The plan's authors identified downtown sites (some with improvements on them) that could accommodate this volume of new space (fig. 5.6). Through these plan documents, the City of Chicago and business leaders signaled to the development community their desire to see additional commercial construction.

More timely and meaningful to real estate investors was the fact that the 2003 plan would become the basis for rewriting the zoning code. The 2004

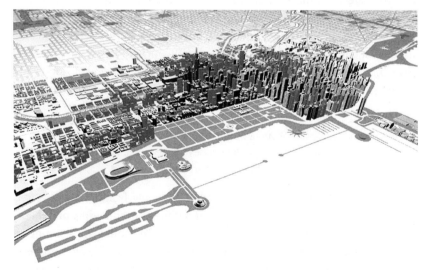

FIGURE 5.6. Rendering of "The Central Area in 2020" for *Chicago Central Area Plan* (2003). Image courtesy of SOM.

rewrite (the first since the zoning ordinance was approved in 1957) was intended to prevent conflicting land uses and provide some certainty to developers as the downtown experienced rapid physical growth. Four new downtown "D" districts in the Central Loop would protect the historic core of office and retail uses with a buffer of mixed-use buildings surrounded by residential areas.[120] The new code also minimized the lot size needed to erect residential uses, encouraging more vertical construction.

The revised code, much like the old one, was only loosely enforced because of a long-standing bonus system that gave downtown developers the right to build out and up in exchange for promises to create public amenities, such as more pedestrian-oriented infrastructure. Triggered by the size and height of a proposed development, this system—known as "Planned Developments," or PDs—allows the city to negotiate the terms of the development and to grant density and use bonuses beyond the base allowances and requirements provided by ordinance.[121] It is through this mechanism that the city bargains for public amenities such as plazas, upper-story setbacks, open space, atriums, and pocket parks, which developers provide in exchange for additional floor area and subsidies. The City Council must approve PDs, and individual elected officials often hold more power in negotiations than planners. The zoning rewrite maintained this ability of City Hall to override regulations and plans on a project-by-project basis.[122]

These tools were used extensively during the Millennial Boom. Of the roughly six hundred large-scale downtown developments during this period, approximately a hundred were PDs. And of these hundred, approximately forty were major office projects, most in the West Loop.[123] Developers tactically increased project sizes so that they could qualify for the PD designation in the first place. In almost all cases the city encouraged larger buildings, but bargained for additional public benefits in exchange for density and height bonuses:

> With little public discussion, Mayor Daley's administration has made a dramatic policy reversal, encouraging great height rather than forcing developers to make their towers shorter. At the House of Daley, where the city's architectural cloth gets cut, tall and thin is in. Short and squat is out.[124]

Thus the city helped push the skyline higher, particularly in the densely built-out core and on sites that appeared to call out for a commanding presence.

If a proposed building adhered to regulations and could win the support of the local alderman (i.e., it did not raise the ire of neighboring property owners), the city would issue the developer a building permit. The Department of Buildings took note of the frenzied pace at which it was permitting new projects, but did so in a detached manner. It did not see its role as managing a potential oversupply through its ability to dole out permits. Planners were "just trying to keep up . . . We didn't have any time to do any real planning."[125] One of Daley's longest-serving planning commissioners noted in the middle of the boom that the city simply "respond[s] to the private sector" and that it guides growth by regulating densities through the zoning process.[126] Beyond such measures, the municipal government did not put the brakes on development even when it did not appear warranted by demand. The city was not comfortable "second guessing" the market, but it was content to encourage that same market to keep generating new assets.

The Asset Assembly Line

No amount of public investment would have spurred the physical expansion of Chicago's downtown had private actors not also been looking to construct and trade new assets. Without heated acquisitions and securities markets, developers and owners would have built and bought much less. They would not have been able to offload their buildings to the next willing buyer, and lenders would have guarded their credit.

Local real estate professionals were aware not only of the debt they owed to capital markets, but also of the problems posed by too much supply. They dealt with its return-dampening effects not by halting development in its tracks, but through less visible ways of managing the overstock. They switched into different building types and submarkets. When the condo market began to show signs of weakening, for example, developers rebranded their projects or turned plans for residential buildings into hotels or student housing.

Less obvious were the ways in which these actors helped construct demand for the new products—sometimes before they were even built. Developers, building managers, and brokers steered residents living in other parts of the Chicago region to downtown condos and townhouses. They helped persuade incumbent tenants to abandon the office buildings erected during previous booms with rent concessions, public subsidies, and talk of obsolete structures. Making a market for the new construction tapped into occupants' preexisting needs and desires while helping to shape and enlarge them. It also involved eliminating some of the surplus at the bottom of the food chain— for example, when the city used TIF funds to encourage the conversion of "obsolete" Class B and C office buildings to other uses. I discuss these and other means of managing oversupply in the following chapter.

Making the Market for Chicago's New Skyscrapers

Village Green Companies, a Michigan-based developer, had already converted two landmarked office buildings in downtown Chicago. The Fisher Building on Dearborn, built in 1896 and designed by D. H. Burnham and Company, stands out for its facade, which features intricate terra-cotta carvings of dolphins, starfish, and mythical sea creatures. It was home to nonprofit organizations and solo practitioner law firms when it was purchased by Village Green, who evicted the tenants and converted the building into 180 apartments and short-term rentals. The Medical and Dental Arts Building on Wabash and Lake Streets was another office-to-residential conversion, this time of a landmark designed by Daniel Burnham Jr.

The principals behind Village Green were third-generation real estate men. They had moved beyond their humble origins as builders to become successful developers, owners, and managers of over forty-two thousand apartments valued at over $2 billion. Their success derived from accepting one trend and fighting another. They knew that a handful of postindustrial midwestern downtowns had outgrown their need for historic Class C office space. Cities such as Minneapolis–St. Paul and Ann Arbor experienced vacancies less because of economic decline than because of competition from new office towers. Class A buildings lured away better tenants, with a ripple effect that left landlords of older ones with the most income-constrained businesses and few economic incentives to renovate. Historic structures were left to deteriorate as ruins among the new glass-skinned towers.

But Village Green bucked another trend: suburban living. They bet that certain downtowns could become residential havens for young cosmopolitans seeking amenities, walkable environments, and proximity to work. Village Green found success in the downtowns of cities that, although not growing, were not declining or suburbanizing as fast as others.

Converting certain Class C offices into its patented "city apartment" product tested the company's resolve. In 2005, when Village Green turned its sights to a forty-five-story Gothic Revival building at 188 West Randolph in Chicago, the structure was crumbling.[1] Built in 1929, the building had been vacant for several years, and its terra-cotta cladding was falling onto pedestrians on the sidewalk below. It had spent several years in expensive litigation as the City of Chicago pursued its former owner for code violations, then attempted to acquire 188 West Randolph outright through eminent domain while the owner was bogged down in bankruptcy proceedings. In 2005 a bankruptcy court approved the building's sale to Village Green for $10 million. But renovation plans for what was eventually to be called the Randolph Tower City Apartments fell through in 2008 when, at the onset of the credit crisis, banks withdrew their backing for the tax-exempt municipal bonds that would have helped finance the conversion. Meanwhile, estimates for construction costs spiraled upward, and the Central Loop Tax Increment Financing (TIF) District, the city's main piggybank, expired.

The aged structure at 188 West Randolph was exactly the kind of building that "the market," if left to its own devices, would chew up and spit out to make way for a modern alternative that took better advantage of its attractive location. "Frankly, this building will be lost without the benefit of public assistance," argued the developer.[2] The city's Department of Planning and Development found a way not only to keep the TIF subsidy, but to increase it to $34 million: it designated a new single-purpose TIF district around the building. In addition, the developer was able to layer on federal and state property and income tax incentives, which provided millions more in public subsidy.

Although some preservationists were pleased to save the building, others attacked the expense to the taxpayer. Why should the City of Chicago assist a developer when private debt was available and trends pointed in the direction of a downtown housing revival? Architecture critic Lynn Becker called the building a "welfare queen" and facetiously congratulated his readers: "You, lucky taxpayer, kick in almost half of the project cost and the private developer gets the building. Socialism, Chicago style."[3]

Socialism is undoubtedly an overstatement. What the Village Green case demonstrates, though, is one way that real estate professionals deal with the oversupply of new construction: the destruction of surpluses left over from previous building booms. Chicago was awash in office space during the Millennial Boom. While some of that surplus could be found in the new office buildings, where tenants were subleasing unoccupied and expensive space, it was mainly their older counterparts that became redundant. The slow-growing workforce, low absorption, and high vacancy rates during this period reveal a market feeding off itself as the new buildings cannibalized older ones. The primary tenant-sending structures and submarkets were stuck at the end of both the vacancy and value chains, saddled with empty, devalued space.

Small developers like Village Green and the City of Chicago stepped in to kill off some of that excess capacity. The city labeled older stock "obsolete" and the submarkets where it was concentrated "blighted." With public money, some obsolete buildings (particularly those with what brokers call "terminal" vacancies) were demolished, or in the case of office-to-residential conversions, the buildings themselves were saved but the office uses were eliminated.

Conversions and demolitions act as a kind of corrective surgery to remove surplus space, tightening markets and allowing them to absorb more of the new stock. These strategies paved the way for more new construction and, by removing cheaper space alternatives, helped push pricing upward (or at least sustained higher prices in the face of oversupply). Improving the vacancy, lease, and absorption rates so important to investors indirectly assisted John Buck, Sam Zell, and other investors to extract more profit from the property base.

Eliminating older buildings also facilitated another strategy for capital accumulation: attracting tenants to new space. With fewer spaces to lease, tenants set their sights on newer buildings. This strategy helped new office developers, for whom signing a lease with a high-status anchor tenant, such as a law firm, was the only way to justify creating new commodities during a period of escalating production. With tenant commitments, new assets could be built and subsequently transacted through the complex chain of acquisitions and debt securitization set in motion after each delivery.

To help make the occupant and investment markets for new construction, developers enlisted brokers to act as go-betweens, negotiating favorable rents and improvement allowances for potential tenants to induce their relocation. Brokers offered persuasive arguments to convince tenants of the shortcomings of their current premises, emphasizing subtle aesthetic distinctions between the old and new buildings that played into tenants' insecurities about their relative social rank. Even municipal governments got in on the game. In the name of "economic development," the City of Chicago offered financial assistance to relocating tenants to ensure that they stayed within municipal boundaries and did not abscond to competing suburbs or cities. This chapter examines the practices and policies that sped tenant movement and helped construct demand during the Millennial Boom.

Musical Chairs

Once the new towers started going up, downtown Chicago experienced a domino effect of leasing activity. The cranes and backhoes, and the noise produced by the builders, made it easy to overlook the moving trucks circumnavigating the Loop, shuttling the possessions of law firms and condo owners to their new quarters. The city resembled, in the words of one broker, "a game of musical chairs."[4]

The number of new office leases signed per year more than doubled between 1997 and 2006.[5] This increase in lease volume occurred despite the fact that tenants typically sign longer-term leases at the top of the real estate cycle because that is when their own business prospects appear most secure. Instead, Chicago office tenants opted for quick deals and extensions, confident that they would not need to lock in rents because rates would not grow quickly.[6] Businesses also chose to sublet space as original tenants took on more than they needed.[7]

Downtown legal, accounting, and financial institutions led the leasing stampede. These were the highly sought-after anchor tenants whose commitments to lease over 100,000 square feet allowed the new office towers to go up in the first place. Law firms accounted for approximately one-third of all major lease deals; during the boom, ten of the twenty largest law firms in Chicago relocated to new quarters (table 6.1).[8] Law firms also occupied the largest share of Loop office space, followed by banks and financial firms, so that every one of their moves sent tsunamis, not ripples, through the downtown market.

Law firms and financial institutions moved to the advancing western frontier of office development. Mapping the origin and destination zip codes of finance, insurance, and real estate (FIRE) establishments whose addresses changed reveals that a small number of submarkets experienced the bulk of receiving activity.[9] Specifically, zip codes in the western periphery of downtown were the destinations for many of the businesses previously located in adjacent zip codes to the east. Those West Loop zip codes with a net gain in business establishments were also those that added the most new rentable building area. In contrast, zip codes in the East and Central Loop submarkets experienced net losses in establishments during the first half of the boom.

As one example, law firm Mayer Brown—with over fourteen hundred attorneys—vacated its home of seventeen years, a postmodern gabled tower

TABLE 6.1. Largest Law Firm Relocations, Chicago, 2003–2012

Firm name & current location (year built)	No. local attorneys	2012 sq. footage	Base rent per sq. ft. per year	Date of most recent move	Former location (year built)	Former sq. footage	Rent per sq. ft. per year prior to move
Kirkland & Ellis 300 N. LaSalle (2009)	615	650,000	$33.00	2009	Aon Center (1974)	700,000	$13.67
Sidley Austin 1 S. Dearborn (2005)	506	575,000	$27.00	2005	One First National Plaza (1969)	376,000	NA
Mayer Brown Hyatt Center (2005)	406	754,076	$28.20	2005	190 S. LaSalle (1987)	453,000	NA
Jenner & Block 353 N. Clark (2009)	310	411,228	$29.95	2009	IBM Building (1971)	358,593	$27.75
Baker & McKenzie 300 E. Randolph	217	237,000	NA	2012	One Prudential Plaza (1955)	256,212	$19.72
Seyfarth Shaw Citadel Center (2003)	215	294,200	$22.11	2006	55 E. Monroe Street (1972)	220,858	$11.16
Skadden Arps Slate Meagher & Flom 155 N. Wacker (2009)	196	223,500	$28.00	2009	333 W. Wacker (1983)	161,300	NA
Foley & Lardner 321 N. Clark (former Quaker Tower) (1987)	140	211,000	NA	2003	IBM Building (1971) / Three First National Plaza	200,000 / NA	$17.70 / NA

Sources: Data from *Crain's Chicago Business* "Crain's List: Chicago's largest law firms," Feb. 21, 2011; press releases; *Chicago Tribune*; CoStar COMPS.

that architect Philip Johnson had designed for it at 190 South LaSalle in 1987. Mayer Brown moved to the new Hyatt Center on South Wacker Drive, three blocks west. Financial firms, accountants, and banking giants such as Goldman Sachs, PriceWaterhouseCoopers, and Citigroup moved to new West Loop towers. Construction of buildings such as the ABN Amro Plaza, a 1.3 million-square-foot tower on the edge of the West Loop, pulled bank offices from their locations on LaSalle Street. Many corporate headquarters also joined the westward migration.

WHY DID THE LAW FIRM CROSS THE ROAD?[10]

At first blush, the relocation frenzy of the 2000s seemed entirely rational and grounded in a profit-maximizing calculus. The decision to stay or go was typically made by a corporation's facilities manager or by a law firm's real estate group, assisted by a team of brokers, architects, and space planners. As providers of due diligence and intermediation services in an increasingly financialized economy, a few law and financial firms experienced hiring jags during the Millennial Boom that exhausted their available space and forced them to consider expansion options. The law firm Sidley Austin, for example, outgrew its home of thirty years and had already relocated some personnel to a second building before moving down the street to a new tower developed by Hines.

The majority of Loop tenants, however, were not expanding in size. Flat employment growth throughout the boom suggests that most inhabitants of new space were no- or low-growth occupants moving from other parts of the downtown. Average lease sizes declined as large corporations consolidated or downsized.[11] In many cases of Chicago law firm relocations, tenants downsized and put more space back on the market than they leased.[12] The trigger for these moves was often an expiring lease. Commercial leases are typically fifteen to twenty years long, and many had been signed when the firms moved into new space in the late 1980s and early 1990s.

Whether firms moved to take advantage of technology, better locations, or other amenities is less certain. The geographic concentration of Class A buildings in the West Loop (more Class B and C buildings were found in the East and Central Loop) and the movement of business establishments to more western zip codes suggest a concomitant improvement in quality for many westward-moving tenants.[13] However, it is difficult to validate claims about whether tenants "do better" after a move to a new building (which would involve chain mapping each major move and comparing the costs and efficiency and productivity gains associated with a firm's new home with those potentially generated by retrofitting the one left behind). The statistical and interview data presented in chapter 4 cast some doubt on the conventional, demand-side explanations for these migratory patterns.

But if demand was weak, why did the inhabitants of Chicago's Loop spend the Millennial Boom participating in an expensive game of musical chairs? Just as developer behavior is conditioned by the supply of capital and facilitated through complex webs of intermediation, tenant behavior also responds to the availability of financing and the pressure of brokering agents. When credit is available to support space consumption beyond available income, businesses and households have an incentive to relocate. Movement is enabled when financing presents few limits. Low interest rates for home mortgages were partly responsible for relocations to downtown condos as households took on record levels of debt to fund home purchases. Similarly, banks were willing to provide commercial loans and lines of credit to fund business operations, a portion of which paid for rent and improvements.

But having more cash on hand does not always lead to greater consumption of space. After all, households could purchase more appliances with their money, and businesses could invest in higher management salaries or jobs. Beyond the satisfaction of basic needs, space consumption is also a social act, and it is therefore malleable by actors on both the demand and supply sides.

The moves were partly motivated by tenants' desire for prestige and status. A partner at Jenner & Block admitted that his firm's relocation depended partly on peer behavior:

> With very few exceptions, all of the big firms in Chicago, they've moved. When we were still at the IBM Building, and our building was just coming up, we started bringing new law students that we were recruiting over to the building to have them look at plans. The other firms that they were going to visit, Kirkland and Sidley, were in new buildings. We wanted to make sure that they knew we were going to be in a new building.[14]

Mainstream analysts identified this dynamic at work in Chicago during the Millennial Boom: "The construction of new buildings is not supported by market demand but rather by the desire of individual companies to anchor their own trophy buildings."[15] Large tenants became more aware of their power to actualize such status-conscious desires as the boom progressed and landlords became more desperate for their occupancy. Noted one law firm partner, "It's unbelievable the amount of capital right now. [Developer John] Buck's probably making $15 to $20 million per building just in developer fees! But they still need anchors like us to seal the deal."[16]

Social conventions ensured that major tenants would steer clear of those buildings that had already been claimed by their competitors. Once one major law firm moved into a new building, other law firms on the same rung of the status hierarchy refused to move there too.[17] That building was symbolically "marked" as the property of the first mover. For reasons of pride and repute, tenants also refused to occupy the space vacated by their competitors, as taking cast-off space was viewed as an admission of lower rank. These conventions fed the desire to relocate and constructed a need for more new buildings that was more social than material.

Developers came up with their plans and acquired sites in advance of their discussions with tenants. But, dependent as they were on pre-sale and pre-lease commitments to secure construction financing, they began to promote their buildings aggressively as soon as they sensed conditions were right. Hines was in conversations to move law firm Sidley Austin across the street from its prior premises four years before the new tower went up: "The building [One South Dearborn] was built because of their occupancy . . . Otherwise we wouldn't have gone forward."[18] In some cases developers persuaded their best tenants to move from older buildings they owned or managed to newer structures that could command premium rents.[19] Developers are "some of the most persuasive salespeople you've ever met. I mean, 'we've got the best product with the best finishes!' And they're not faking it: they believe their own stories."[20]

Despite these powers, office developers budgeted an estimated 10 percent of their project costs for marketing expenses, a figure that could easily exceed $10 million for one building.[21] These dollars went to the city's top brokerage firms, which provided the interface between tenants and developers, building owners, and investors.

In Cook County, approximately 4,700 people were employed as commercial brokers in 2007.[22] Acting as tenant and owner representatives, a select group of them were at the eye of the whirlwind of short-distance office relocations downtown: national firms CB Richard Ellis, Studley, Cushman and Wakefield, Jones Lang LaSalle, Colliers, and Staubach, as well as Chicago-based Draper & Kramer and Baird & Warner, two of the city's oldest and largest brokerage firms.[23] Many of these firms can trace their lineage back to the Chicago Real Estate Board, which was founded in 1883 when the city was rebuilding from the Great Chicago Fire. As in other cities, the Chicago board set professional standards and hosted social events for members. But more so than other city boards, Chicago's was deeply involved in civic affairs and boosterism. From underwriting the bid for the 1893 World's Fair to fomenting interest in nuisance laws, zoning, and bond caps, Chicago real estate brokers have been a powerful force in city policy making.[24]

Partly because of its habitual oversupply, Chicago is known as a "tenant rep-driven market," whereas in other cities, such as Manhattan and San Francisco, landlords call the shots.[25] Major tenants typically have exclusive relationships with one brokerage firm. Noted one facilities manager at a large law firm, "We had a long relationship with Colliers and they helped us out a lot during this whole process. They beat the streets and got proposals from JBC [John Buck], Prime . . . They handled all of the information coming into our firm."[26] Chicago's real estate brokers moved in the same social milieu as developers and anchor tenants, and in the words of one consultant, they "fed off each other's confidence and adrenaline."[27] "It was like a college football game here," noted another, with teams made up of developers and their brokers picking off tenants from their opponents and then "high-fiving."[28]

In addition to the motive of solidifying their own social status and that of their clients, brokers had monetary incentives to relocate businesses to developers' new buildings, as their compensation depended on their ability to shift tenants to more expensive quarters. Individual brokers and their firms can make more from new buildings because lease rates there are higher. Brokering is a commission-based business, and every time a tenant relocates, the representatives involved draw handsome fees from building owners—typically between $1 and $2 per rentable square foot, or 3 to 5 percent of a year's rent.[29] At the peak of the 1980s boom, a standard commission of 30 percent of the first year's rent for a ten-year lease (or 3 percent annually) could exceed $1 million for a 100,000-square-foot deal.[30] At that time, building owners also offered Porsches, vacations, and cash bonuses to brokers who could help them ink deals. During the Millennial Boom bonuses became less showy but no less valuable. As the oversupply of space downtown increased, developers and landlords offered financial rewards to tenant representatives to position their buildings at the top of the listings. When building owners were flush with cash from fresh sales, they upped the ante and offered larger cuts of the annual lease price to brokers who could fill their buildings.[31]

During the Millennial Boom Chicago brokers marketed several of the new office buildings as "Class A+," a designation they accorded to those with the most popular addresses and the highest-status tenants and architects. They emphasized the exclusivity of the new stock while denigrating the old. Brokers peppered their proposals with the language of obsolescence, underscoring the ways in which tenants' older buildings were failing to fulfill occupants' space requirements and how retrofitting would be more expensive than relocating. One building owner suggested that brokers weighted the scales in favor of new buildings from the onset of lease negotiations: "Brokers don't present tenants with a real alternative to see what would happen if they stayed. They're trained to do the simplest kind of lease analysis, and they always seem to make the new building come out ahead . . . In a sense, the real estate industry has created that feeling of obsolescence."[32]

Success ultimately hinged on brokers' ability to get their clients the best deals and negotiate favorable lease terms. Said one tenant representative, "Our job is basically to shuffle tenants around from one building to the next. The pitch is: if you move, we'll save you money."[33] Brokers slashed effective rents by extracting concessions from office building owners.[34] Concessions, which at other times were considered one-time marketing specials, became institutionalized during the overbuilt Millennial Boom.[35] Typical concessions included periods of free rent; generous improvement packages to finish a tenant's space; cash rebates to help pay for moving expenses; lease "pick-ups" in which the new landlord would pay to release a tenant from its prior obligations; and on occasion, equity positions in a new development.[36] One year's rent was the standard abatement period during the boom (one month for every year of the lease), with larger concessions for larger tenants.[37] While free-rent periods stayed constant over the boom, other concessions, such as tenant improvement allowances, increased over time, peaking in 2006–2007 as vacant space piled up.[38]

When concessions are taken into account, relocations to what looks on paper like more expensive space appear more economically rational. Noted one broker, "If you're finishing up a fifteen-year lease where rents have been escalating by two-and-a-half percent a year, you can afford to move. With concessions, you may not have to pay more to lease new space."[39]

Despite their ubiquity, concessions were not reflected in the published data on lease rates (in other words, "effective rents" remained elusive). Brokers guarded this information for fear of its getting into the hands of their competitors and other landlords; some leases even stipulated that information on concessions would not be released to the public.[40] Such evasiveness served three purposes. First, it contributed to the impression of value provided by the brokerage industry in an age when more data were publicly available. Second, it kept tenants from learning that their neighbors got a better deal than they did, confronting their landlords for their own concessions, or feeling bitter and leaving. Third, it allowed valuations and published indicators of the supply-demand balance to look better than the reality. Most reports show Class A asking rents in downtown Chicago increasing by about 30 percent between 2004 and 2008, despite the fact that brokers knew such growth was offset by generous rent abatement packages and improvement allowances.[41] Inflated published rents for Class A buildings created yet another signal for developers to increase inventory at times of weakening demand.

Competing landlords were forced to retaliate by lowering their rents to

hold onto tenants and attract new ones as brokers negotiated aggressive coun-
teroffers. When the Chicago office market went into a slump in 2001, REITs
such as Equity Office Properties drastically cut asking rents, and "property
managers were told they would lose their jobs if they lost an existing tenant
to a competing building."[42] EOP also doubled broker commissions on lease
renewals.[43] This strategy may have worked for individual landlords (occupancy rates in EOP buildings surpassed industry averages during this time),
but concessions weaken the market as a whole by giving the illusion that it
can support additional supply.

ECONOMIC DEVELOPMENT OR
REARRANGING DECK CHAIRS?

The concessions offered by landlords of new buildings might have been
enough to induce tenant moves, but City Hall also jumped into the fray. Chicago has a long history of subsidizing peripatetic corporations. Most of its
business attraction incentives, however, involved establishments moving to
Chicago from a location *outside* of the city. In a classic case of "bidding for
business," for example, the City of Chicago and the state of Illinois provided a
package of incentives worth $56 million to Boeing to relocate its headquarters
from Seattle in 2001. Emboldened by its victory over Dallas and Denver, the
city began extending TIF subsidies to businesses *already located in* Chicago.
Municipal planners justified their use of TIF as a necessary intervention to
prevent existing headquarters from moving out of the city.[44]

The City of Chicago approved TIF allocations ranging from $1 million to
$27 million to move corporate headquarters from one downtown building
to another. The city provided TIF assistance to at least fourteen corporations
headquartered downtown during this period (table 6.2). These corporations
ranged from major players in the finance (ABN Amro, CNA Group), food
and beverage production (Quaker, MillerCoors, Barry Callebaut), manufacturing (Mittal Steel), and transportation (United Airlines) sectors to small
tech start-ups (NanoInk).[45] The allocations typically took the form of allowances for moving expenses and tenant build-outs of the new space.

Approximately one-third of these subsidized relocations were from older
to newer (built after 2001) buildings, as in the case of US Gypsum (USG), a
corporation that makes drywall and other building materials. USG moved
its headquarters from the USG Building at 125 South Franklin (completed
during the previous boom in 1989) and entered into a fifteen-year lease in a
tower that local developer Fifield completed in 2006 at 550 West Adams. USG
received almost $10 million in TIF funds from the city to assist with its move.

TABLE 6.2. TIF-Assisted Headquarters Relocations, 2000–2011

	Date	TIF assistance ($)	Total project cost (million $)	Percent TIF assistance	Expected jobs retained or relocated	New location (year built)	Prior location(s) (year built)
Quaker	10/30/2000	9,750,000	89.5	11%	800	555 W. Monroe St. (2001)	321 N. Clark St. (1987)
NanoInk	7/1/2003	1,000,000	5.9	17%	4	1335 W. Randolph (1928)	1335 W. Randolph (1928)
ABN Amro*	10/26/2003	27,037,699	342.4	8%	2,700	540 W. Madison St. (2003)	135 S. LaSalle St. (1934)
USG	11/8/2004	9,750,000	128.2	8%	500	550 W. Adams St. (2006)	125 S. Franklin St. (1992)
CNA Group*	11/2/2006	13,680,000	63.7	21%	3,200	333 S. Wabash Ave. (1973)	333 S. Wabash Ave. (1973)
UAL Corp. (United Airlines)	8/8/2007	5,475,000	20.9	26%	365	77 W. Wacker Dr. (1992)	1200 E. Algonquin Rd. Elk Grove Village, IL (1974)
Infinium	12/11/2007	1,377,451	11.7	12%	300	900 N. Kingsbury St. (1930)	Chicago Board of Trade 141 W. Jackson Blvd. (1930)
Ziegler	3/10/2008	2,416,000	8.3	29%	151	200 S. Wacker Dr. (1981)	Milwaukee HQ; 1 S. Wacker Dr. (NA, 1982)
Mittal Steel	3/13/2008	2,000,000	6.5	31%	212	1 S. Dearborn St. (2005)	Various offices in Ohio, Indiana, and elsewhere (NA)
Crate & Barrel	10/14/2008	3,400,000	16.2	21%	72	240 N. Ashland Ave. (1929)	1250 Techny Rd. Northbrook, IL (HQ) 1201 W. Washington St., Chicago, IL (Photo Studio) (2001)
CareerBuilder	11/5/2008	2,900,000	11.7	25%	1,600	200 N. LaSalle St. (1984)	200 N. LaSalle St. 8420 W. Bryn Mawr 180 N. LaSalle St. 111 N. Canal St. All located in Chicago, IL (1984, 1981, 1971, 1913)

(continued)

TABLE 6.2. (*continued*)

	Date	TIF assistance ($)	Total project cost (million $)	Percent TIF assistance	Expected jobs retained or relocated	New location (year built)	Prior location(s) (year built)
MillerCoors	9/9/2009	5,775,000	21.5	27%	325	250 S. Wacker Dr. (1957)	Miller: Milwaukee HQ Coors: Denver HQ (NA)
UAL	11/19/2009	24,389,768	71.8	34%	2,500	233 S. Wacker Dr. (1973)	1200 E. Algonquin Rd. Elk Grove Village, IL (1974)
NAVTEQ	1/21/2010	5,000,000	28.6	17%	900	100 N. Riverside Plaza (1990)	Merchandise Mart 222 Merchandise Mart Plaza (1930)
Barry Callebaut**	11/18/2010	880,000	3.5	25%	80	600 W. Chicago Ave. (1908)	Quebec HQ (NA)
Chicago Mercantile Exchange*	12/8/2010	15,000,000	65.0	23%	2,388	141 W. Jackson Blvd. 10 S. Wacker Dr. 20 S. Wacker Dr. 30 S. Wacker Dr. 550 W. Washington Blvd. All located in Chicago, IL (1930, 1987, 1929, 1983, 2000)	142 W. Jackson Blvd. 10 S. Wacker Dr. 20 S. Wacker Dr. 30 S. Wacker Dr. 550 W. Washington Blvd. All located in Chicago, IL (1930, 1987, 1929, 1983, 2000)
Sara Lee	12/13/2011	6,500,000	30.1	22%	650	400 S. Jefferson (1946)	3500 Lacey Rd. Downers Grove, IL (1992)

Source: Data from City of Chicago TIF Redevelopment Agreements.

*TIF contributions were fully or partially returned to the City.

**Redevelopment Agreement unavailable.

TIF dollars can be viewed not only as benefiting the tenant, but also as covering a portion of the developer's overall building costs. After word of the USG relocation subsidy spread, some downtown real estate firms complained to the Daley administration that the subsidies were fueling new development at a time when vacancy rates were already too high.[46] But both city and developer repudiated the idea that public policy had anything to do with overbuilding. Echoing John Buck's portrayal of developers as humble servant to tenants' desires, Steve Fifield countered: "Development is dictated by the big tenants. It always has been."[47] The city ignored Fifield's competitors' request to limit the practice and continued to allocate property tax dollars for micro-moves downtown. The developer went on to sell its new building to a German investment fund for approximately $168 million, generating an estimated profit of $45 million.[48]

Economic development practices like these can be viewed as enabling, or at least validating, the choices of tenants to relocate from older structures to slightly newer ones. Although city planners never stated that this was their goal, and in some instances expressed concern about the effect of tenant movement on older buildings, they too had reasons for filling the new buildings.[49] Newer Class A buildings, particularly those with high occupancy, had higher assessed values, and therefore generated more property taxes per square foot, than comparable older buildings. Commenting on the USG deal, the city's spokesman said the new tower would generate $3.9 million annually in property taxes, whereas the vacant lot on which the tower was built had provided only $131,000.[50] Moreover, the city had already made concentrated infrastructure investments in the submarkets where new construction was located. "Building more office space in the West Loop is part of the goals of the *Central Area Plan*," said the city's spokesman.[51]

The mad rush of tenants relocating to new buildings, some with city assistance and some with private concessions, strengthened the market for the new towers. With the help of brokers and city planners, tenant interest could be massaged out of a weak economy, which gave developers and their financial backers the green light to keep building.

Creative Destruction: Renovation, Conversion and Demolition

LOOP HOLES

The supply shock of new buildings catalyzed a series of moves until some landlords and submarkets were left with fewer occupants when the music stopped. When tenants are mobile, what happens to one building at one point in space and time redounds to affect others. Specifically, in a no-growth situation, any relocation will leave a comparable unit vacant, unless it can be filled by another occupant making a move or is removed.

Tenants occasionally switched ranks between the three different classes of office buildings. For every year of the Millennial Boom (with the exception of 2000), Class B and C buildings in the East and Central Loop experienced higher vacancy rates and lower absorption than Class A construction (fig. 6.1; see also fig. 4.13), suggesting some upward mobility. In a few isolated years during the cycle, particularly those immediately following a rash of building, Class A landlords provided enough concessions on rent so that a few Class B tenants were able to move into higher-quality space.

However, tenants were constrained by income and space requirements that made major upgrading, or what is called "filtering" in the housing literature, difficult.[52] Comparing vacancy rates for Class B and C buildings with those for their Class A equivalents shows a narrowing between 2003 and 2010, suggesting that tenants were not all leapfrogging from lower to higher ranks.[53] Similarly, the difference in gross asking rents between Class A and B space never appears small enough to allow trading up en masse to be feasible.[54] Class B and C tenants generally take less space, and new Class A buildings were not willing to lease at lower volumes (many maintained that a full floor or, if pressed, a half floor was the minimum), holding out instead for large-block tenants.

Both prior research and my interviews confirm that horizontal churning was more common than vertical filtering.[55] Tenants generally remained within the same rank in the hierarchy; that is, tenants moved from Class A buildings from the 1970s and 1980s to Class A buildings constructed in the 2000s. In the case of such lateral moves, the existence of filtering is difficult to confirm. Moving from, say, a building from the first Chicago skyscraper boom of the 1920s to a glass-and-steel tower circa 2005 might occasion a major upgrade. But the difference in amenities between that same 2005 tower and one constructed twenty years earlier is likely to be much smaller.

The cannibalization occurring downtown was no secret. One reporter observed, "Those Super Class A or A Plus buildings just sucked tenants out of older buildings in the East and Central Loop. It was a zero-sum game; the chess pieces just moved around the board."[56] Appraisal Research Counselors, one of the leading market analysts in the city, concurred: "While much of the new construction space was pre-leased, there are now holes created in the downtown market due to vacating of previously occupied space. There has been limited new demand for space from corporations outside the downtown

Average vacancy rates

○ 15% or lower
● higher than 15%

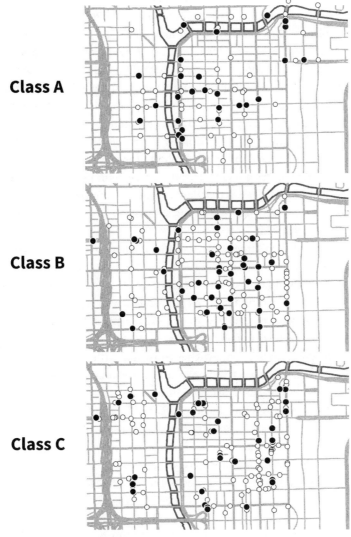

FIGURE 6.1. Office buildings with high average vacancy rates, East, Central, and West Loop submarkets, 2006 to 2009, by class. Data from CoStar COMPS.

market. Rather, companies are relocating *from older, obsolete space to more ef-*
ficient, new space in the market" [italics mine].[57]

Real estate professionals downplayed the facts that high vacancies in older
buildings were related to the new stock and that the vacated premises, still
serviceable or convertible to contemporary office design standards, were
made obsolete by the proliferation of new space that attracted their tenants
away. Instead, they stigmatized the losers as *inherently* flawed and deserving of
their fates. They were "dogs"; "They're old and they look it"; "They are over-
finished; I mean just look at all that marble and fancy woodwork"; "They look
fake with all that veneer"; "Their floor plans are all wrong"; "They don't have
enough contiguous space"; "These buildings weren't even that competitive to
begin with. They were put up so quickly in the 1980s."[58] One of John Buck's
brokers disparaged a tenant's prior building for giving off a strong smell of
the 1980s: "I remember going over and meeting with them [law firm Locke,
Lord Bissell & Liddell]. It was that dark ruby red or cranberry carpet and dark
walnut on walnut with more walnut."[59] In many ways, they blamed older
buildings for being old.

Certainly older buildings started out with a handicap. Class B structures,
for example, were short (averaging sixteen stories as opposed to thirty-eight
floors for Class A buildings), and their floor plates, on average, were smaller
than those in Class A buildings.[60] As in most cities, the age of Chicago buildings was negatively correlated with acquisition prices.[61] It is important to
note, however, that prices tend to decline faster in the early years of a building's life than they do later on, which is consistent with Dermisi and McDonald's finding that "vintage" Chicago buildings maintained their values better
over time than recent-past construction.[62]

Indeed, the buildings hurt most by overbuilding and the westward exodus
of tenants during the Millennial Boom were not the oldest specimens, but
rather the products of the two prior booms: the 1960s and the 1980s. Class A
buildings in the West Loop and Class B buildings in the Central Loop built
during these booms posted the highest volumes of direct vacant space, almost
3 million square feet each in 2008.[63] Office buildings constructed only twenty
to thirty years earlier "sucked wind . . . after their anchor tenants moved to
new buildings," noted one tenant, and could not be "back-filled" by new
businesses.[64] Modernist towers such as One Prudential Plaza (1955), 11 East
Adams (1961), and the IBM Building (1971), as well as more recent icons such
as the diamond-topped Smurfit-Stone Building (1984), had hit or were near-
ing "the end of their thirty-year life cycle . . . They were fully depreciated
but not yet appreciated."[65] Class A buildings that only fifteen years earlier
were "top-of-the-market trophies" were so hard hit by tenant moves that

some questioned whether they could still be considered Class A or should be downgraded instead.[66] These postwar towers found themselves in a kind of purgatory: they could not lower their rents sufficiently to hold onto tenants and still turn a profit, but neither were they historic or cheap enough to be converted to residential use.[67]

Financial markets contributed to the obsolescence of these recent-past buildings in ways other than by underwriting the new buildings that were their direct competitors. Office buildings generally need to undergo a major retrofit every eight to ten years in order to maintain their Class A status.[68] Some buildings missed out on their retrofits because their short-term owners, such as Sam Zell, could profit more from flipping them than from reinvesting in them. Architect Frank Gehry, a minority investor in a consortium that purchased the sleek Inland Steel Building (1956) in 2005, famously griped that the landmark's renovation was put on hold while its owners held out for the highest acquisition price.[69] Meanwhile, the architectural icon was at least 50 percent vacant.

The use of CMBSs to acquire buildings encouraged overleverage and underinvestment. Even before the crisis hit, many CMBS-financed acquisitions were sinking underwater financially, as they were appraised at inflated values and their rental income, even with high occupancy, could not carry the debt service. Lenders prevented owners from modifying leases and reducing rents to hold onto anchor tenants. Buildings such as the 10 and 30 South Wacker Drive complexes (1983; 1987) and 161 North Clark (1992) had lost occupants before Tishman Speyer purchased them from Sam Zell's Equity Office Properties. Tishman could not unload these towers fast enough to avoid plummeting rents and appraisals. When they were worth less than they were purchased for and owners could not refinance their mortgages, these buildings became "zombies"—occupied and alive, but just barely.[70]

Owners of those Chicago buildings vacated by the chain reactions of moves had limited options. Forget the major renovations needed to reposition a building—even basic maintenance was difficult when operating revenues were restricted by high vacancies, low rent growth, and punishing debt. For those buildings experiencing cash-flow problems because of overleverage, financial distress quickly became physical distress when reinvestment ground to a halt. In 2010, for example, the forty-six-story office tower at 500 West Monroe (built in 1991 by Tishman Speyer) became the first foreclosure of a major Loop office in eleven years and a "sign that the market remain[ed] mired in the hangover of the debt-stoked valuation bubble that peaked in mid-2007."[71] The building had been purchased in 2007 by New York–based Broadway Partners for $340 million, using a package of loans that were securitized and made up more than 95 percent of the purchase price. Three years later it was worth only $240 million, it was 30 percent vacant, and renovations had been postponed.

Owners that had borrowed big to develop or purchase buildings with shrinking occupancy occasionally defaulted on their loans. Of the 129 downtown condo projects on the market in 2008, 53 ran afoul of their lenders.[72] Between 2008 and 2011, about 60 commercial buildings in the Central Area entered foreclosure proceedings.[73] Of the hundred largest U.S. metro areas, Chicago had the highest delinquency rate on commercial real estate loans.[74]

Not surprisingly, most of the office foreclosures were 1980s buildings located in the East and Central Loop submarkets, and many had been purchased with CMBS financing.[75] Because commercial mortgages are typically nonrecourse loans (i.e., landlords are not personally responsible if they default) and borrowers are prevented from declaring bankruptcy to extricate themselves from troubled financial arrangements, some owners of underwater commercial properties just walked away and forfeited their assets.[76] Rather than restructuring complex securitized loans with scores of bondholders, owners sent the buildings' keys to their lenders—a phenomenon acerbically known as "jingle mail."

RENOVATION

Options for absorbing excess capacity generally fall into two general categories: renovation or removal. Developers can revalorize zombie buildings—particularly if they have historic or architectural merit—by reinvesting in their existing uses, by converting them to other uses ("adaptive reuse"), or by demolishing them and building new. The uncertain results of such strategies, the small cohort of developers pursuing them, and the slow pace at which they unfold invite public sector intervention to expedite the creative destruction process. Local governments can reduce the friction of market adjustment and safeguard demand for the new supply.

In Chicago, reinvestment was limited to those buildings that were sold to new owners with deep pockets and manageable debt. For example, the Insurance Exchange Building at 175 West Jackson (built in 1912 and 1928) was a Class B building purchased by Intell from Helmsley-Spear in 1999. The new owner transformed it into a Class A building by redesigning communal spaces and adding an indoor parking lot as well as new windows, elevators, and wiring. The architect Lucien Lagrange described it as "a very postmodern building. The outside will keep its early 20th century look, but as you enter the building, you enter a very contemporary world." This discrepancy was appealing to the young technology companies that moved in.[77] A handful of recent-past skyscrapers with name brands and architectural pedigrees underwent similar multimillion-dollar renovations. Chase Tower (formerly the First National Bank of Chicago Building and then Bank One Plaza; 1969), the Harris Bank Building (1974), Madison Plaza (1982) at 200 West Madison, and 190 South LaSalle (1987) all lost major tenants to Wacker Drive during the Millennial Boom, becoming up to 60 percent vacant.[78] The new owners decided to undertake major modernization schemes, including new lobbies, new HVAC and elevator systems, fitness centers, and even private clubs in the effort to offer amenities akin to those provided by their competitors on Wacker Drive. But even the new owners of retrofitted skyscrapers admitted that the renovations would not translate into rent premiums and that they would "probably only be able to compete with low As and high Bs."[79]

Because they knew that the city maintained an interest in preserving the most beloved examples of historic architecture, property owners made entreaties for public assistance. The LaSalle Central Redevelopment Project Area (fig. 6.2) was a response to such requests and embodies many of the goals and

FIGURE 6.2. LaSalle Central TIF District. Data from City of Chicago Department of Housing and Economic Development.

contradictions of public involvement in booms.[80] This TIF district was designated just as the boom was evolving into a bubble in 2006, and it targeted one
of the submarkets hardest hit by the relocations to new construction on the
Loop's western periphery (the other being the East Loop, some of which was
covered by the Central Loop TIF district). This intervention was intended to
help an area with deficient infrastructure and "outmoded" office buildings
compete with the West Loop—specifically, "to provide resources for their
rehabilitation for current and new uses, especially projects that involve the
district's numerous historic structures."[81]

In order to classify the area as worthy of public funds, the city first had to
demonstrate, in the middle of a construction boom, that its buzzing downtown was "blighted" and would not be developed "but for" the use of TIF.
The consultant's eligibility report found sufficient evidence of blight, mainly
by documenting that 63 percent of buildings in the district were over thirty-
five years old (the age that, not coincidentally, early appraisal treatises considered to be a building's natural life span).

Although the TIF redevelopment plan remarks that "the excessive vacancy rates appear to be linked to the migration of many firms to new office buildings on the periphery of the Central Business District," it defines
obsolescence as an innate attribute of structures that "make the market reject
[them]."[82] The stock of older buildings, it states, "is facing high vacancy rates
due to their obsolescence" [italics mine]. If a district's buildings had vacancy
rates above 20 percent for five of the previous ten years, if net rents were less
than $5 per square foot, or if property values had not risen as fast as they
had for the city as a whole in four of the five years previous to the study,
they "exhibited" obsolescence.[83] Additional evidence included office towers
that "no longer fit the changing demands of tenants," such those designed
with interior load-bearing walls and columns, dated mechanical systems, and
small floor plates.

After the LaSalle Central TIF district was designated, however, few of the
obsolete buildings in the district received TIF funds for renovation. Other
than routine infrastructure improvements, the city paid mainly for relocation incentives so that tenants could move from other parts of the downtown
(mostly from the equally traumatized East Loop) to the core.

The irony of boom-time public assistance and the slippery category of
blight did not go unchallenged. Community organizations such as the Grassroots Collaborative built a campaign to shut down the LaSalle Central TIF
district, making an example of the largesse directed at building owners and
corporate tenants. Protesters at one point tried to present one of the subsidized firms, the Chicago Mercantile Exchange Group, with a golden toilet

meant to symbolize the millions of dollars in TIF funds allocated to renovate the Board of Trade's bathrooms. Although the CME Group and a few
other recipients elected to return their TIF subsidies (CNA Group, Bank of
America, and United Airlines admitted that incentives were no longer necessary to their continued operations), the TIF district remains in place until
2029—sufficient time to catch the next wave of speculation, appreciation,
and redevelopment.

CONVERSION

Surplus office buildings that were considered historically or architecturally
distinguished were converted to residential condos, apartments, dormitories,
and hotels—many with TIF assistance. Loft conversions downtown had occurred in a spotty manner since the 1970s, but garnered attention only when
they took place in the core, as did architect Harry Weese's 1981 conversion
of the Manhattan Building (1891) on Dearborn. The 1996 conversion of the
Singer Building set off a later trend, by the end of which twenty-seven downtown office buildings had been adapted to residential uses.[84] Except for four
of these, the conversions took place in the East and Central Loop submarkets.
These areas were not only the best suited for residential uses (with narrow
streets and proximity to amenities such as Millennium Park), but also contained the largest concentration of pre–World War II building stock.

The conversion trend was driven by a combination of market behavior
and city policy. The developers drawn to these projects (e.g., Village Green,
Marc Realty, Concord, Smithfield, Centrum, and Belgravia) were headquartered in the region and were smaller than the office developers on Wacker
Drive. They encountered several roadblocks along the way. First, approximately thirty Class C office buildings were being held in a kind of redevelopment limbo, as they had been recommended for landmark status, but the
City Council had never passed the requisite ordinance to protect them.[85] Second, conversions were expensive. Because of the increasing value of downtown locations, average acquisition costs for Class B and C office buildings
increased from $30 per square foot in 1998 to $125 in 2006, not including the
expense of relocating tenants or buying out their leases.[86] Third, lenders were
more circumspect about adaptive reuse projects than they were about new
construction, requiring padded reserve funds and additional equity for such
projects.

Yet the City of Chicago maintained a strong interest in seeing these projects to fruition. Office-to-residential conversions allowed the city to achieve
several goals. Most immediately, they preserved the ornamented facades of

buildings that gestured at the Loop's history and gave the downtown the visual diversity and texture that could promote a 24-hour experience and tourism. Indirectly, conversions allowed for the continued construction of high-end office towers that could accommodate the corporate headquarters and business service functions of a global city. Removing some of the surplus Class B and C buildings from office uses relieved some of the pressure on the flagging office market by creating a situation of contrived scarcity.

In contrast to New York City's Lower Manhattan Revitalization Plan, which offered a package of zoning changes, financial incentives, and rebates on utility costs to encourage conversions, the Daley administration never announced a comprehensive initiative to adapt surplus offices to residential uses.[87] As with many of its urban development policies, the City of Chicago pursued its interests in conversions in a less coordinated manner that stayed mainly below the radar. The city was inconsistently supportive of conversions as it tried to reconcile the divergent interests of tenants, developers, and building owners.

When one of the buildings on the list of "pending" landmarks was demolished in 1996, the city moved in to protect the others. In exchange for limiting landlords' ability to demolish their buildings, however, the city convinced Cook County to offer property tax abatements to even non-landmarked historic buildings.[88] The abatements created an incentive for redevelopment but did not protect a building's prior use.[89] In almost all cases, landlords that took advantage of this new "Class L" abatement for historic properties converted their Class B and C office buildings to residential uses.

In addition to the tax abatements, the city's other tool for jump-starting conversion activity was, once again, TIF. Alderman Burton Natarus (who chaired the Plan Commission and sat on the powerful Zoning Committee) cited the expense of office-to-residential conversions as the main justification for the North Loop TIF district. Public monies would be needed, aldermen claimed, to upgrade older buildings' electrical and sprinkler systems and to protect the glazed terra-cotta cladding and metal filigree on their facades.

Twenty of the twenty-seven major conversions in the Loop received TIF assistance. Most of these conversions occurred in the early years of the boom, with the private market taking over once the viability of these experiments was established. Many of the buildings that received TIF funding were designed by architects associated in some way with Daniel Burnham, the patron saint of Chicago architecture. Perhaps the best-known case of a City-sponsored rescue was the Burnham Hotel. Built by John Root and Charles Atwood as the Reliance Building, this fifteen-story office building on State Street was converted into a boutique hotel at the beginning of the boom in

1999. The city not only contributed millions in TIF assistance, but also purchased the building before selling it to Kimpton Hotels. Other office-to-hotel conversions followed, including that of the black granite-clad Carbide and Carbon Building designed by Burnham's sons, which was turned into a Hard Rock Hotel.[90] National developers such as American Invsco and Draper and Kramer jumped in, and projects, such as the 220-unit Metropolitan Tower redeveloped by Daley ally Louis D'Angelo, got larger. In the East Loop, offices also were converted into student dormitories or classroom space.

Of the twenty office buildings to receive TIF assistance, only two maintained their office uses. This observation suggests that preserving physical facades was a higher priority for the city than protecting land uses. Losing out were the small tenants that occupied historic Class C office space: mostly nonprofits, start-ups, and law and medical practices. They were subject to rapid rent increases, (sometimes illegal) lease terminations, and eviction.[91] Displacement compromised their ability to locate near their clients, volunteer bases, and transit. In many cases these tenants ended up being displaced again when the buildings to which they relocated were themselves converted or upgraded for higher-income tenants.[92] The utility older office buildings provided was seen as a small price to pay as these uses were sacrificed on the altar of progress.

DEMOLITION

Most of the Class C office buildings converted to other uses during the 2000s certainly would have been demolished in previous booms. The last great wave of downtown demolition occurred during the overbuilding of the 1980s, when even buildings designed by beloved architect Louis Sullivan were flattened to make way for surface parking lots and new office towers along Wacker Drive.

Downtown demolitions were relatively rare occurrences during the Millennial Boom. Approximately ninety demolition permits were issued between 1996 and 2007, most of them for small structures. Not only is it physically difficult to demolish skyscrapers, but by the twenty-first century the value of historic preservation was more deeply embedded in the practices of the local government and the consciousness of voters. Still aghast at having watched some monumental structures succumb to the wrecking ball in the 1980s, the city enacted policy reforms that protected the remaining historic stock. In 1987, for example, the Chicago City Council increased the city's power to prevent the demolition of designated landmarks. When Richard M. Daley came into office, he was amenable to a preservation-tinged agenda; the mayor, after all, "liked pretty things."[93] Over half of Chicago's 277 (as of 2009) landmark designations occurred under his watch.

Despite Daley's concern for preservation, he also habitually yielded to pressure from developers—as evidenced by the twenty-five years it took to pass the ordinance designating the Michigan Avenue Landmark District and by the fact that over 4 percent of the buildings on the Chicago Historic Resources Survey (a color-coded priority system for the city's historic inventory) were demolished, in several cases with no imminent redevelopment. Those downtown buildings that did come down were primarily located in the Central, North, and East Loop submarkets.

The former home of the Chicago Mercantile Exchange, a seventeen-story building on the corner of Franklin and Washington, was one of the major casualties of the Millennial Boom (fig. 6.3). Built in 1927, the building had large arched windows and stone panels carved in bas-relief depicting hens, eggs, fowl, and merchants on its limestone exterior. The interior was decorated with gold-leafed molding and white marble and contained elaborately gilded elevators. A new owner saw little reason to maintain this Beaux Arts building, which was notably not landmarked and not Chicago School. In 2003 the building was razed. The ease with which the developer was able to obtain a demolition permit roused preservation advocates into action, prompting a series of passionate protests in period dress. While the site remained vacant for several years after the demolition because of financing problems, the protesters successfully lobbied the city to adopt an ordinance requiring it to wait ninety days before issuing a demolition permit.

Unlike that of the Mercantile Exchange, most demolitions were part of executed redevelopment plans that left the sites only temporarily empty.[94] The International-style Sun-Times headquarters came down in 2005 so that the Trump International Hotel and Tower could rise from its ashes. Few would miss the Sun-Times Building. Erected in 1957, this squat seven-story rectangle had flat sides and a lid. The building's boxlike form was, to quote an editorial celebrating it, "versatile and efficient, working as well for office employees and printing presses as it does for shoes and dead people."[95]

The John Buck Company built 111 South Wacker, a fifty-one-story blue-glass office tower, on the former site of the US Gypsum Building (1963), one of the tallest and most modern buildings to be demolished in Chicago. A century-old office building that had once served as Citibank's Loop headquarters was demolished to clear the site for Hines's metal-and-glass tower and plaza at 1 South Dearborn.

During the boom the North Loop—home to the city's Magnificent Mile of shopping—lost many of its remaining historic assets. New retail develop-

FIGURE 6.3. Chicago Mercantile Exchange before (*top*) and after demolition (*bottom*), 2003. Photographs courtesy of Landmarks Illinois.

ment marched south on Michigan Avenue, taking out buildings such as the Arts Club of Chicago, which was designed by Mies van der Rohe. Opposition was focused on the noteworthy interior of the club but was not sufficient to convince John Buck to adjust his plans for a low-rise retail box. Embarrassed by the incident and by the developer's heavy-handed efforts to quell opposition to his plans, the Daley administration pushed back on Buck's next project, requiring JBC to dismantle and reconstruct the exterior of the

McGraw-Hill Building (1929) for its new shopping center. This directive required JBC and its financial partner, Morgan Stanley, to pay an estimated 10 percent more to redevelop the site.[96]

The city not only permitted demolitions, but in several cases committed TIF funds to help pay for building removal. Demolition was among the most popular "eligible uses" of TIF funds because it reset a parcel's value downward and allowed the future increment generated as part of the redevelopment to be appropriated entirely by the TIF district. For example, the narrow 174 Randolph (1875), once the headquarters of the Showman's League of America, was demolished to make way for a pocket park for John Buck's 155 North Wacker tower. Because the site lay just outside the edge of the Central Loop TIF district, it was initially ineligible for subsidy. The city, however, rewrote the LaSalle Central TIF district boundaries to include the parcel, and TIF funds were used for demolition and streetscaping improvements that enhanced the value of Buck's new building.[97]

From Bubble to Rubble

Putting up new space in a no- or low-growth context sets off a chain reaction, a sequence of moves that occurs as a new unit is claimed by a first mover, which leaves a vacant unit behind, which is claimed by a second entity that leaves another vacant unit behind, and so on. In Chicago this reaction to the new stock was fueled less by Pavlovian responses to more cost-effective space than by status, incentives, and intermediaries whose livelihoods were inexorably linked to the building boom. Tenant relocation activity was directed toward hot spots of new construction and away from older submarkets.

The exhilarating hypermobility of the Millennium Boom gave the appearance of progress and growth. But this restlessness took place in a city with flatlining employment, population, and rents. Herein lies one of the most perverse dynamics of building booms: that pumping up markets for new construction sends signals to developers to add even more supply. The additional supply hurts many of the actors whose behavior encouraged its construction in the first place. When the bubble burst, investors experienced losses, developers could not complete their projects, brokers and appraisers lost their jobs, and the City of Chicago's fiscal stress worsened.

Some individual actors survived the bust because of good luck or timing. If they could minimize their own financial exposure, find an anchor tenant, and wiggle out of their ownership positions shortly after their buildings were erected, developers kept building. As one observer noted, "They [developers] only want to fill their own buildings. The rest of the market could fall into Lake Michigan for all they care."[98]

At the other end of this bifurcated market sat landlords of buildings left over from prior booms. The new office towers along Wacker Drive poached from the older Class A and B buildings in the Central and East Loops, whose owners struggled with vacancies that threatened the already precarious financial arrangements that had enabled their purchase. Overleveraged yet under-leased, fully depreciated yet underappreciated, these buildings fell into decline and, in some cases, foreclosure. The office towers from the 1960s and 1980s booms were hard to convert, with acquisition costs that were too high and floor plans and finishes that did not lend themselves to adaptive reuse. If they could not secure the label "modernist" or lacked the name appeal of a famous architect, these recent-past buildings hemorrhaged tenants and spiraled downward, with little hope of being transformed into boutique hotels or high-end historic apartments. Unless the economy grows, downtown offices of this vintage are sitting ducks for demolition or high-stakes conversions when the next tidal wave of capital hits.

PART THREE

Building the Future

7

The Slow Build

The slow, the old, the small in ambition, the devotee of the outmoded have no less right to property than have the quick, the young, the aggressive, and the modernistic or futuristic.

MORRIS V. DISTRICT OF COLUMBIA REDEVELOPMENT
LAND AGENCY 1953, 38[1]

So far I have focused on the systemic causes of overbuilding, particularly the historically and locally specific practices through which abstract processes of capital circulation and accumulation assume physical form in the production of buildings, submarkets, and urban landscapes. But what are the typical effects of overbuilding? What kinds of spaces and places does it tend to produce? Are overbuilt cities more or less interesting, more or less economically and environmentally sustainable? Who benefits from the whirlwinds of construction that occur during booms and bubbles, and who gets stuck with the costs of overly ambitious physical growth?

Most real estate economists would consider such questions imprudent. To believers in efficient markets, overbuilding is just a short-term aberration— one that should not be judged too quickly, if at all. Market rationality will prevail in the long run, with prices oscillating but eventually adjusting to the changed supply-demand conditions. The high vacancy rates associated with overbuilding will disappear as prices are reset and excess space is either discounted and absorbed by less affluent occupants or eliminated though demolition and conversion.

Putting aside the question of whether markets really are efficient, one problem with adopting this wait-it-out approach is that the incentives for overbuilding are never identified or addressed, and its effects are just kicked down the road. Decisions to add new buildings may induce public costs or trigger negative externalities (witness the hulking ruins of empty big-box stores and subdivisions dotting some geographies). Moreover, insisting that markets are efficient in the long run provides little guidance to those actors who do not possess the luxury of time when determining whether they are facing a short-term predicament (such that overcapacity will prove beneficial

in the future) or whether they will be stuck in states of excess supply for significantly longer spells.

This chapter describes the near-term effects of overbuilding. It also lays
out possible policy solutions to overbuilding that would diminish exposure
to its more damaging consequences and encourage the modernization of existing building inventories. The case for regulation does not stem from a belief in the aesthetic superiority of older buildings or sentimental attachments
to the status quo. Rather, curbing overbuilding and reusing existing space
is the more fiscally, economically, and environmentally sound approach to
managing urban change, one that can accommodate growth and innovation
while also valuing attachments to the artifacts and landscapes of the past.
Urban planner Kevin Lynch wrote that "an environment that cannot change
invites its own destruction," yet he also noted that places that grow slowly
are often richer, more complex environments with choices and services better suited to the plurality of needs and values of a diverse population.[2] "Slow
cities" grow in a deliberate manner that values quality of life, the preservation
of difference and environment, and place attachment.[3] Slow, smart cities are
preferable to rapid and unsustainable urbanization.

The Benefits and Costs of Overbuilding

IN PRAISE OF EXCESS

The push for faster, more dynamic cities is led by those who measure their
success in terms of groundbreaking and ribbon-cutting ceremonies. Developers and investors who follow the proven formula—build, attract tenants,
and sell while the market is hot—do very well during the boom-time crush.
If their timing is right, they can avoid being saddled with onerous debt service payments or carrying vacant units on their books. If they profit during
the boom and bubble, they can ride out the ensuing bust and start all over
again once conditions improve. Other development service providers also
welcome overbuilding. Neither the surveyor nor the concrete pourer will
break a sweat if the buildings they helped erect remain empty for some time.
During periods of rapid construction, their workloads spike, their employees
are well remunerated, and their contracts are honored. And even when new
buildings are poorly occupied, the tax revenues and pride they generate give
local governments no incentive to stem the flow of permits.

Periods of overbuilding are more than just a busy time for architects
and engineers; they are defining moments for these professionals as cities
are turned into canvases for their ideas. Sophisticated and fantastical designs

typically relegated to renderings on a computer screen can be given material form when money is flowing and corporate clients are taking risks with their commissions. During the Millennial Boom modern architecture pushed the limits of what was buildable with structures that were more sustainable (e.g., generating their own energy), transparent (with light, exposed skeletal supports), and intelligent (wired to provide and respond to real-time information) than any before them.[4] The undulating, wind-resistant balconies of Jeanne Gang's Aqua (see fig. 4.8), the poetic massing of Frank Gehry's band shell in Millennium Park, and the glass curtain and parking garage integration into the lobby design at Goettsch's 111 South Wacker are but a few examples of innovative design in Chicago during the Millennial Boom.

In Chicago as well as in cities across the world, the new crop of skyscrapers was taller, yet less massive and bulky, than the buildings that preceded them.[5] A "tall building" craze proceeded with the support of a global architectural movement dedicated to the pursuit of ever greater heights and of scholars who provided that movement with intellectual justification. Economist Edward Glaeser, for example, makes the case that greater scale is the key to accommodating growth and depressing prices for everyone: "Simply put, the places that are expensive don't build a lot, and the places that build a lot aren't expensive . . . Growth, not height restrictions and a fixed building stock, keeps space affordable and ensures that poorer people and less profitable firms can stay and help a thriving city remain successful and diverse."[6] He attributes the relative affordability of Chicago's housing stock to a more permissive regulatory environment than in, for example, New York City, which has restricted development in prime locations. Ignoring the fact that Chicago posted much weaker employment, income, and population growth than New York City over the same decade, he concludes that "the forest of cranes along Lake Michigan keeps Chicago affordable."[7]

Overbuilding may also be the temporary articulation of new geographies of agglomeration and innovation, the benefits of which could be more obvious in the long term. Construction booms in creative hubs such as Seattle and Austin, even if those booms overproduced, could signal an economic shift toward younger businesses, workers, and cities.[8] Similarly, overbuilding in a downtown may make sense given the attraction cities hold for a more youthful demographic and start-ups that are likely to grow. From a planning perspective, a regional structure that includes a dense core is highly desirable because it takes advantage of prior infrastructure investments, such as those in transit systems, and diminishes the tendency to sprawl. During the Millennial Boom Chicago was one of a handful of metropolitan regions where the majority of office space was located in its downtown.[9] Observed one informant, "I know [we overbuilt] but businesses are attracted to the Loop because of the great space here. You don't see that in a Des Moines or Kansas City. Those same businesses would have gone straight to some flex space in the suburbs."[10] Partly due to the availability of space and reasonable rents, Chicago's downtown outperformed its competition in the suburbs, where office parks and off-ramp campuses were even more under-inhabited.[11]

An empty office tower is still viewed as a symbol of wealth and progress. It plays an important marketing role for those image-conscious members of the growth machine whose status and viability depend on maintaining a high level of construction activity. Its vacancy is often invisible to those forming opinions about the prosperity and promise of a city and is less important than the virtue of change represented by the building itself. The conviction that the constant renewal of a city is socially beneficial was ever present among Chicago real estate professionals as they equated scale and novelty with progress.[12] For them, overbuilding was a fleeting ailment, not a chronic disease, and they were confident that its symptoms would be alleviated by demolitions in the short run and by inevitable economic growth in the future.[13]

PRETTY VACANT

Overproduction can be beneficial if a surplus of new construction sets off a process of filtering. The housing filtering model, popularized in the 1940s and 1950s, argues that new, better-quality space catalyzes a chain reaction of moves that eventually forces property values downward, at which point vacated units become available to households with lower incomes. In other words, the cascades of relocations that follow episodes of overbuilding can be welfare-enhancing if the newly occupied units are always of a higher quality or lower price. If occupants all "trade up" and their former premises are leased by others (which happens in a growing economy), overbuilding will not be a problem.

Many observers interpreted the spate of office overbuilding in Chicago during the Millennial Boom according to this model. Noted one, "A continuous stream of new state-of-the-art space is undeniably a very healthy trend for downtown, as space is freed up in older buildings (often at quite reasonable asking rental rates) to attract new and expanding businesses to constantly re-shape and strengthen the downtown economy."[14] Unfortunately, as I have shown in chapters 4 and 6, there was little empirical evidence to substantiate such claims during the 2000s. Prices were sticky and commercial markets were highly segmented; if occupants moved, they tended to move to new units in the same real estate class.[15]

Even after a series of lateral moves, however, the "worst" buildings within each market segment are left behind. What happens to such buildings? Some are demolished or undergo conversions to new uses. But not all obsolete buildings are destroyed. As industrial designer George Nelson noted, "The vacuum cleaner obsoleted the broom, but people still use both."[16] Buildings that are economically obsolete are still habitable, and if their rents decline in tandem with their perceived quality, they may attract a different, more income-constrained class of users.

Nontraditional occupants gravitate toward obsolete spaces, which are affordable and display the grit that many purveyors of alternative cultures find glamorous and exciting. Examining transitions in Chicago's Wicker Park neighborhood during the 1990s, sociologist Richard Lloyd describes the area's transformation, which required "neither a massive clearing project bulldozing material relics of industry nor a wholesale erasure of local history."[17] Instead, the apartment buildings appealed to a young, hip, asset-poor demographic of artists and businesses. Similarly, expanding technology companies such as Google, Braintree, and Sandbox moved into renovated loft-style warehouses and Class C offices in the far West Loop. These kinds of buildings proudly exhibit their distinctiveness through signifiers such as off-the-beaten-path locations and chicly tarnished finishes (e.g., exposed brick). Older buildings and salvaged materials have become markers of taste for this substratum of the creative class.

Obsolete spaces also become sites of adaptive reuse. In addition to the Class B and C office buildings converted into hotels and apartments (see chap. 6), other building types vacated in previous rounds of industrial restructuring were appropriated for new uses during the Millennial Boom. Warehouses in the West Loop were transformed first into artists' lofts, then into higher-end living spaces for professionals and students, and then into retail and office space.[18] Pop-up stores and art galleries appeared on the ground floors of vacant apartment buildings in the South Loop.

Even in poor, forgotten places, small-scale but spirited types of reuse thrive—as evidenced by the photo essays of Camilo Jose Vergara,[19] media artist Julia Christenson's Big Box Reuse project,[20] and Theaster Gates's installations.[21] These artists demonstrate how the dilapidated, devalued spaces of capitalist urbanization—vacant Victorian houses in Detroit, abandoned strip malls outside of Kansas City, foreclosed commercial buildings on the South Side of Chicago—can provide the seedbeds where alternative uses and do-it-yourself cultures can flourish. Cultural practices like these often mobilize irony in ways that reconcile some of the conflicting attitudes about obsolescence and modernity, loss and change, what is kept and what it thrown away.

While these kinds of ironic, quirky spaces make cities distinct and diverse places, such projects are likely to be one-offs or temporary uses.[22] Successful rehabilitation and adaptive reuse projects are difficult to pull off when rents must be low enough for tenants without access to credit to afford. If older buildings with little apparent historical or architectural cachet attract any investor interest, it is likely to be small and local money. These strategies may also unwittingly pave the way for more organized investment circuits, mainstream consumer culture, and demolition in their environs.[23]

COLLATERAL ENVIRONMENTAL DAMAGE

There may be reasons to welcome the occasional bout of overbuilding, particularly if it brings imaginative design, more efficient structures, and opportunities for income-constrained tenants to inhabit higher-quality, second-hand space. However, the costs associated with overbuilding can be high and are usually borne by weaker actors at scales far removed from the instigators and beneficiaries of the boom.

For one thing, building a new structure when an existing one would have served the same purpose is wasteful. Construction consumes tremendous amounts of natural resources and energy: gypsum is mined in Oklahoma for drywall, forest products are harvested in Canada, and China provides U.S. contractors with tons of stainless steel. As raw materials and finished products are being harvested and manufactured in increasingly faraway locales, they must be shipped longer distances, depending on the locations of their points of origin and consumption. Long-distance trips increase fossil fuel use and carbon emissions and also raise prices for consumers.

The construction process itself creates great quantities of waste, which localities must process, dump, or export. The bulk of local waste streams is comprised of materials produced by building activity; approximately 160 million tons of construction and demolition debris is generated in the country every year, enough to erect a wall 30 feet high and 30 feet thick around the entire coast of the continental United States.[24] These materials include scrap lumber and other wood products, brick and block, gypsum wallboard, manufactured wood, asphalt shingles and pavement, metals, plastics, concrete, and dirt, as well as salvageable appliances, ornaments, and fixtures.

An estimated 60 percent of the total volume of construction and demolition debris is sent to landfills, where it increases the level of spatially concentrated toxins that may leak into the ground or become airborne emissions.[25] Landfills and unpermitted dump sites are often located in minority neighborhoods. In Chicago, for example, the 80-foot "mountain" of illegally dumped construction and demolition waste in the low-income, predominately African-American neighborhood of North Lawndale for years symbolized the environmental degradation and corruption endured by the city's poorest neighborhoods as a result of growth occurring elsewhere in the region.[26]

In contrast, reusing a commodity maintains its "embodied energy" and creates less pollution, all while keeping its component parts out of the waste stream. Although the new buildings that replace older ones may be more energy efficient than their predecessors, most engineers and architects still agree with the old saw that "the 'greenest' building is the one not built."[27]

SPECULATING ON THE PUBLIC'S DIME

Overbuilt landscapes can become their own kind of pollution or unwanted land use, "like refuse or scrap cars."[28] As zombie subdivisions, empty office parks, and vacant shopping malls attest, even new buildings may go uninhabited for years—especially when their ownership status is complicated by foreclosures. In the United States, housing units built since 2000 had the highest vacancy rates at the time of the bust.[29] In some western counties 15 percent to two-thirds of the lots entitled for subdivisions were still vacant six years after the crash.[30]

Such persistent vacancies may be due to the fact that new construction is not always as innovative or well located as growth advocates believe. Despite the opportunity to make great leaps forward in architecture, engineering, and city planning during the Millennial Boom, much of new commercial construction was generic and uninspired. Even the gleaming new towers on Chicago's Wacker Drive did not win over all critics. Some called them "anonymous" and "commodity buildings, architectural fill . . . structures of little distinction that serve their limited purpose."[31]

Some even questioned the quality of the new structures relative to the older buildings they were replacing or, at least, outnumbering. In Chicago, neighborhood three-flats were hastily erected by small, inexperienced builders to cash in on what they knew was a terminal boom. Even downtown condo developers "cut corners" by the end of the boom as they struggled to keep up with demand and to find competent contractors who were not committed to other jobs.[32] Although not condoning these shortcuts, capital markets were neither encouraging attention to detail nor promoting sustainable designs.[33] If their purchases were overleveraged, little cash flow was available to fund maintenance and upgrades in the ensuing years. Some buildings erected during the Millennial Boom will probably be considered obsolete when the next boom occurs.

Most new construction during the Millennial Boom occurred not in cities at all, but in far-flung suburbs (what Robert Lang and Jennifer LeFurgy call "boomburbs").[34] Several of the worst offenders in terms of office overbuilding—Atlanta, northern Virginia, Phoenix—were also those metropolitan regions with the greatest tendency to sprawl.[35] One study estimated that 5 to 10 million excess single-family homes on large lots were located on the edges of metropolitan areas in places with little infrastructure.[36] Low-density construction creep consumes more natural resources and requires more expensive infrastructure to service than comparable development in urban locations.

Overbuilt markets are not always cheaper ones. Contradicting the laws of supply and demand, several of the regions where the production of housing units was higher than the national average, such as Las Vegas, also experienced the greatest valuation bubbles.[37] Moreover, several office markets with shares of new construction and vacancy rate increases exceeding the national average, such as Washington, DC, sustained higher-than-average office rents throughout and immediately following the boom.[38] Economists suggest this is caused by "stickiness," a resistance to sharp downward price movements.[39] Instead of slashing them, property owners fix prices below which they will not sell, holding out hope for higher bids and anticipating strong future appreciation.

When the occupancy of new buildings comes at the expense of older buildings, the vacated buildings become less fit to meet contemporary requirements, such as energy efficiency or interior design flexibility. Such buildings may languish and sit idle for years, opening themselves up to nuisance or illegal uses if they are not properly managed. The public sector is called in to respond to concerns for community health, safety, and welfare; local governments may have to pay to acquire and demolish buildings that are abandoned or present immediate or future hazards, such as fire-related damage or excessive code violations. And if buildings do come down, they do so with an emotional power rarely felt for ordinary consumer items. Obsolete buildings are often beloved despite their infirmities; they are important touchstones that stimulate memory and emotion.[40]

From the public sector's perspective, overbuilding also incurs costs by both exaggerating and deconcentrating demand for scarce collective resources. The introduction of new buildings in a static market may disperse the same number of tenants over a larger area, lowering densities and resulting in the underutilization of available public resources.[41] Local governments must cut roads and provide fire and police protection to buildings that are half empty. They must construct schools in uninhabited boomburbs to accommodate enrollments that never materialize, funded by bonds whose repayment schemes involved development charges and impact fees that are never paid. When infrastructure is extended haphazardly, it raises the cost per unit of development served. Local governments must absorb these costs (or pass them on to taxpayers) without the benefit of the additional property taxes.

Following in the footsteps of bullish developers, planners provide entitlements and "overzone" for uses that are not justified by market demand, which can result in long-lasting imbalances of land uses.[42] The tendency of developers to build quickly to excess—and then not to build again for years—keeps the relationship between public planning and private development cycles chronically asynchronous. Fast cities confound planning not because the public sector is by nature inept, but because it follows the lead of private actors who are either ignorant or indifferent about the social costs of their behavior.

MACROECONOMIC INSTABILITY

The costs of overbuilding also circulate nationally and globally when such episodes trigger subsequent financial crises. After the Millennial Boom the United States experienced its worst economic decline since the Great Depression. The shock from the country's residential subprime market collapse propagated to the commercial sector and then on to other countries that had been incubating their own real estate bubbles. By 2009, with a few rare exceptions (Sweden, Australia), the world was facing a financial meltdown of epic proportions.

Because of the growing integration of real estate and finance, most episodes of crisis in the twentieth century have been preceded by boom-bubble-bust patterns in real estate construction and pricing. Overbuilding and speculation alone, however, are not enough to catalyze a crisis. If builders and buyers use their own money, they can generally ride out the turbulence. In the worst cases, they suffer losses when they sell their assets in a depressed market. Building booms in which borrowing is restrained tend to deflate with fewer major economic disruptions, which is why economists are generally less interested in overbuilding than in how it is financed.[43]

It is an entirely different situation when all actors are leveraged to the hilt, as they were during the Millennial Boom. Strong demand for structured finance instruments compelled the issuance of more loans, so borrowers could put little of their own cash into constructing and acquiring property. Historic rates of property appreciation were surpassed by the growth in mortgage

debt. With the progressive detachment of borrowers from originating lenders, risk assignments became "increasingly unclear and incentives for due diligence worsened, leading to insufficient monitoring by loan originators and
an emphasis on boosting volumes to generate fees."[44] In contrast to previous
busts, delinquencies shot up *before* the economic downturn began, revealing
a prerecession disjuncture between debt and borrower or building income.

As a result of its overbuilding, the United States experienced a sharper
fall in property values overall than many other countries, even before the
supply of credit tightened.[45] Commercial property prices then fell 50 percent between their 2007 peak and 2011.[46] The balance sheets of developers,
investors, and their creditors imploded, and troubled assets went back to
lenders or ended up in the hands of special servicers. Uncertainty about the
quality of the underlying loans caused the secondary and tertiary markets
in which these derivatives were transacted to collapse. Once investors were
scared off, a reinforcing cycle began as deleveraging forced a mass liquida-
tion of assets.[47]

The damage wrought by the overhang of property and these souring financial arrangements extended far beyond the real estate industry. In the
United States, real estate was the most important vehicle for storing and ac-
cumulating wealth.[48] The decline in housing prices following the boom left
one-third of households underwater on their mortgages and destroyed an
estimated $4 trillion of household wealth.[49]

The federal government had to come to the rescue of many of the affected
actors, including the banks that had helped to orchestrate the crisis by buying
and selling asset-backed securities.[50] Over 90 percent of all commercial bank
holdings of CMBSs were concentrated on the balance sheets of the twelve
investment banks that received Troubled Asset Recovery Program (TARP)
funds. State governments provided unemployment and welfare benefits not
only to workers in the construction industry, but to others laid off during
the bust. From the municipal level on upward, the public sector fired its own
employees and dramatically decreased its budgets.

The glut of bad debt and vacant property slowed the recovery. Recessions that coincide with a real estate price deflation are generally deeper and
last longer than those that do not.[51] Banks, which have yet to claw their way
out from under the volume of bad loans they made during the boom, have
been reluctant to lend to even the most sure-footed projects. Businesses,
households, and developers are suspended in a time warp as they wait for
the cycle to begin again. Ironically, experts and amateurs alike argue that
more construction and new financial instruments are the only way to climb

out of the economic decline and financial crisis caused by overbuilding and overleverage.

Keeping Overbuilding in Check

The effects of overbuilding are mixed and reverberate at different scales. Reducing the amplitude of building cycles through regulation would address a major source of urban and macroeconomic instability, but it might also squelch some of the creative energy and diversity of form found in dynamic cities, not to mention its effects on the tens of thousands of individuals and businesses that depend on the construction industry for employment and income. Is it worth trying to address overbuilding through policy and planning?

The response from most real estate professionals is a resounding no. Historically, the development industry has either promoted market-driven solutions (such as demolishing the older buildings made obsolete by their newer replacements) or advocated for urban policies that benefit the private sector (subsidizing developers to demolish the same buildings).[52] Through lobbying associations such as the National Association of Home Builders, professionals have pressed the federal government to keep interest rates low, to maintain and extend tax benefits for property ownership and construction, and to allow commercial banks to transact securities.[53] From the industry's vantage point, overbuilding is less of a problem than obstructionist public policies, such as height restrictions and preservation ordinances, that keep developers from maximizing profits.

The chronic incidence of overbuilding, however, shows how "developers and lenders abdicate their responsibility to exercise stewardship over the real estate market" and misallocate public and private resources.[54] In the face of the public costs that arise from what is individually rational yet collectively irresponsible behavior, state intervention can be justified. Developers have few incentives to change their behavior in the absence of taxes or penalties that force them to internalize these costs.

What kinds of policy prescriptions could potentially curb the worst excesses of overproduction without inflicting major damage on local economies? If the public sector wishes to apply the brakes to new construction, it can administer incentives or apply regulatory friction to (1) increase the cost of new construction; (2) lower the expected return from new construction; or (3) extend the time horizons of developer-investors. Such reforms target both the symptom (overbuilding) and two of its causes (the financial instruments that lead to excessive leverage and the professional practices that privilege new construction). These measures are less focused on mitigating the effects of the resulting busts.

Growth Controls

The industry has created its own trip wires to force developers to establish market demand for their products before moving forward with construction projects. The underwriting criteria of banks are intended to be the strongest line of self-defense against overbuilding. In addition, the rising costs of materials and labor are an endogenous drag on new construction. But these brakes are insufficient in the face of capital gluts and moral hazard situations. Creditors have incentives to override such strictures and increase loan production, even when unwarranted.

In recent years some local governments have attempted to wrest authority for balancing supply and demand away from private actors. Under the broad rubric of "growth controls," they have adopted policies that either restrict development overall or confine it to certain places (sometimes called "growth boundaries" or "urban containment zones").[55] Most development approval processes require applicants to meet mandatory building and zoning codes and height limits. On top of these controls, some local governments also require developers to demonstrate that they will not cause excessive harm to the environment or tax the capacity of available public facilities. If they do, they may be required to pay fees to cover new infrastructure or environmental remedies.

While adding costs to any developer's budget indirectly limits production, other regulations directly restrict the number of units or square footage that can be produced.[56] The growth boundary of Portland, Oregon, for example, encumbers development at the city's periphery while streamlining the permitting process and allowing higher densities within the boundary. San Diego assesses stiff charges to projects outside targeted urban areas. More stringent and complex regulatory processes add costs, uncertainty, or delays to development. They have been shown to slow construction starts, reduce the amount of developable land, and restrict the number of builders active in the markets they cover.[57]

Growth control policies are common on the West Coast and in states undergoing rapid suburbanization (such as Maryland). Most have been instituted to protect farmland, prevent environmental degradation, and limit the expense of providing dispersed infrastructure.[58] They have not been designed to reduce overbuilding. In fact, these regulations may have a limited effect on total supply if developers try to compensate for lost opportunities on the fringe by overbuilding within designated growth boundaries.[59] Increasing allowable densities in the core, as well as the general uncertainty surrounding permitting in growth-managed regions, could result in larger buildings downtown. Such policies would not address the prospect of excessive vacancies, as determining whether a new project is marketable is still left to a developer's best judgment.

To confront overbuilding directly, local governments would need to more actively regulate the timing and scale of development, not just its location. For example, they could institute development moratoriums or quotas to condition the issuance of building permits on a schedule. Boulder, Colorado, restricts the total number of new housing units constructed per year. In other countries the public sector acquires land and releases it to private developers in phases when it believes the timing is right, as happens in the United States only to protect open space or fragile ecosystems. As part of their decision about whether to dispose of publicly owned land or issue building permits, local governments could require that applicants demonstrate market demand—much in the same way as banks (are supposed to) do.[60]

Approving permits based on a market's point in the real estate cycle would require local governments to more carefully monitor demographic patterns, demolition activity, price movements, and the new supply pipeline. If they saw early signs of overproduction (e.g., vacancy rates rising above a certain threshold), the public sector could pull back. At a minimum, local governments could limit the subsidies provided to developers during booms. Indeed, the worst time to stimulate development is when the cycle starts to climb; public incentives then make the crests higher and the crashes more catastrophic.[61]

Unfortunately, most municipalities lack the capacity for such planning or, like the City of Chicago, cannot be bothered to adjust permissions and subsidies to market conditions. Asked about the prospect of regulating permits, one Chicago administrator responded, "Why would *our* market analyses be any better than the ones developers commission?"[62] In such cases, a higher-level authority, such as a county, regional planning agency, or state government, may be in the best position to administer or impose these requirements.[63]

Historic Preservation

In addition to tightening underwriting standards, the other way in which the private sector polices overproduction has been to swiftly eliminate the resulting surpluses. While on the surface it is a sensible means of reducing supply, demolition of older stock does not address the threat posed by new construction to natural resource protection or place attachment. The real estate industry tends to write off resistance to the loss of older buildings as sentimental and subjective, failing to recognize that historic structures contribute to the appeal of urban living. Attachments to these structures can be converted to a monetary value if need be—as indicated by the numerous studies that have found a positive correlation between historic preservation and enhanced property values,[64] positive economic outcomes,[65] and high willingness to pay.[66]

An alternative way of discouraging overbuilding would be to encourage the modernization of older buildings instead of the construction of new ones. The shells of buildings could be preserved (as their "outsides" are public and more often the source of distinction and collective memory) while interior modifications would allow them to deliver amenities comparable to those of new buildings.[67] Modernization enables existing buildings to house current, active uses and to generate sustainable cash flows.[68] If developers were more interested in rehabilitating buildings with the latest systems and technologies, they would have less motivation to build from scratch, especially during periods of lackluster absorption.

The physical potential of most commercial buildings is almost infinite; architects agree that most spaces can "be made to suit any function within certain volumetric limits."[69] Preservation can deliver higher returns to developers, as the costs of renovating older structures are lower than for new construction. Some older buildings may even surpass newer ones in terms of energy efficiency.[70] If public benefits such as neighborhood identity and visual coherence are taken into account, the economic case for preserving older structures is even stronger.[71]

Tenants appear willing to pay market rents for renovated buildings, particularly when their employees are drawn from a younger demographic. In Chicago, one developer estimated that rents in top-of-the-line vintage renovations, such as the Wrigley Building on North Michigan Avenue, were comparable to those in the lower reaches of the Class A segment.[72] Decked out with all the modern conveniences and promoted aggressively by brokers, these buildings can set a new standard for what is considered desirable, much as the penchant of technology firms for rehabbed loft space has done in Chicago's Near West Side submarket.

But still, large developers do not tend to view modernization favorably. They have to work within an existing template as opposed to creating their own. Their financial backers are skeptical. Their architects are challenged by the need for customized and idiosyncratic solutions to problems that in new buildings are standardized. Reconstructing everything from the cornices to the light fixtures requires the services of expensive specialists. Building codes are unsympathetic, and brokers are unfamiliar with the product.[73]

When the relative margin for financial feasibility is small, the role of the public sector is critical. After the federal historic tax credit was scaled back in 1986, many states and local governments started providing below-market loans as well as tax deductions, credits, and abatements to encourage preservation. In addition to permit fee waivers and facade rebates, some historic buildings in Chicago are eligible for the Class Landmark ("L") tax abatement, which freezes property taxes at the pre-redevelopment level. This benefit made many of the office-to-residential conversions in the East Loop during the Millennial Boom financially feasible. Seattle and Boston have passed historic district ordinances that limit demolition in particular areas, compel standards of upkeep, and set up review boards to scrutinize development proposals. Local governments can also lease their own space in renovated historic buildings rather than commissioning new ones.

Preservation policies will be ineffective, however, if the other dynamics propelling new construction remain intact. If cheap land is readily available, for instance, developers will simply bypass historic areas and build in unprotected ones where "greenfield" sites are available. In the face of finance-driven overbuilding, historic preservation efforts are small remedies occurring at the tail end of a chain of vacancies motivated by factors over which local advocates and planners have little control. Rather than spending their time convincing a handful of developers to renovate to exact historical standards, preservation advocates should focus on preventing obsolescence in advance of the decision to demolish. They would do well to move beyond site fights over individual buildings and esteemed architects to educate themselves about the drivers of new construction that result in specific submarkets and structures becoming threatened in the first place.

To make these relationships more visible, municipalities could require that new-build developers conduct impact studies for all construction over a certain size. Instead of measuring environmental impacts, such a study would evaluate the building's effect on existing comparable structures in the same or neighboring submarkets. It would answer the question: If the expected tenant base is local, from which buildings would they be likely to move? If the municipality determined that a new building would cause substantial vacancies in other buildings, it could levy a linkage or impact fee that would be available for renovations, land use conversions, or demolitions in the affected submarkets. In this way the cost of the externalities triggered by overbuilding would be "borne by the stream of users rather than by the latest heir."[74]

DECREASE NET INCOME FROM NEW BUILDINGS

Monetary Controls

Local planning policies can influence the location, use, and physical form of new deliveries, but on their own, such policies cannot restrain waves of construction. The price of asset-backed securities set in global markets is a greater motivation for new office towers than municipal density bonuses are. Capital supply is influenced more by macroeconomic policies than by local conditions and plans.

How should monetary policy deal with asset price bubbles and overproduction? Increasing interest rates would elevate the cost of borrowing and slow the supply pipeline by reducing demand for mortgages. Monetary tightening would force sale prices to descend from the lofty perches reached during bubbles. Even if insufficient to curb overproduction, reducing leverage in the financial sector would make the busts less painful, as banks would have fewer bad debts from which to recover.

The Federal Reserve helped engineer the last global real estate bubble by slashing interest rates to historic lows. Central banks like the Fed could therefore help prevent the next one—if they knew when to intervene and how to distinguish lending sprees justified by growth from those set off by speculation. By monitoring conditions, they could adjust interest rates to the different phases of the cycle, stabilizing construction swings and capital flows generally. Central banks in Australia and Sweden moved to increase interest rates when prices began to escalate (2002–2008 in Australia and in Sweden from 2005–2008). Although housing prices there increased, appreciation was muted, and both countries recovered from the busts more quickly than their peers that did not engage in monetary tightening.[75] In 2011 China deliberately engineered a credit crunch by raising interest rates and down payment requirements.[76]

Financial Regulation

Raising the risk standards of the banking and securities industry would reduce the amount of credit going to projects of dubious market interest.

For example, if the federal government required banks to hold more of the mortgages they originate, they would be more focused on loan applicants' estimates of occupant demand and future income. Higher capital adequacy ratios (the proportion of a bank's equity to its assets) and reserves, limitations on the use of riskier loan products, and ceilings on portfolio exposure to real estate would prevent overleverage and could help banks build a buffer against bust-period losses.

Several of these ideas were taken up in the Dodd-Frank Wall Street Reform and Consumer Protection Act of 2010. Dodd-Frank implemented rules intended to prevent the worst of the transgressions that led to the 2008 financial crisis. Risk retention requirements, specifically the requirement to retain 5 percent of all loans, force banks to spend more time screening loans and attesting to asset quality. These rules affect how CMBS loans are priced and will translate into a higher cost of lending, particularly for higher-risk borrowers.[77]

While this framework makes laudable headway toward curtailing incentives for overleverage, it leaves the basic securitization system intact. Industry lobbying intensified as the act progressed through Congress and successfully defanged its most stringent provisions.[78] Weakened by infighting, the federal government has not increased the budgets of the implementing agencies such as the SEC (the agency's chair estimated it would need to hire eight hundred employees to apply the new rules).[79] While CMBSs may now be underwritten more carefully, the absence of a far-reaching overhaul of the derivatives market means that risk taking will reemerge as the market normalizes.

Altering the federal income tax code to provide fewer incentives for new construction would encounter the same political obstructions as financial regulation, but could also exert a powerful check on overbuilding. Most national tax systems provide favorable treatment to property owners with leveraged assets through incentives such as depreciation and mortgage interest deductions. Diminishing the value of these forms of tax relief and limiting other opportunities for tax avoidance on real estate–generated income could dampen appreciation, especially for those investors subject to high marginal income tax rates.

Increasing local property tax rates can moderate building booms when those rates are capitalized into the price of property and suppress its value.[80] Most municipalities are wont to use taxes as a countercyclical tool; they typically lower local tax rates when values are climbing—as Chicago did during the Millennial Boom. But rather than relying on opaque economic development tools such as tax increment financing to surreptitiously increase spending while holding tax rates low, cities need to concede that while booms can be revenue-enhancing, bubbles are expensive. Tempering appreciation, maintaining affordability, and managing development could all justify boosting tax rates during periods of heightened speculation.

INCREASE HOLDING PERIODS

Investors tend to focus myopically on the short-term performance of their buildings rather than surveying future user needs or market conditions. When assets are flipped, they become someone else's problem. Their future is of little concern to their owners, who acquire the capital to move on to the next project. This deeply engrained short-termism puts pressure on building and asset managers to forgo those improvements with high fixed costs and long-run payoffs, whereas cash infusions from sales make strong, immediate impressions on the investment market.

If regulatory carrots and sticks could extend the time horizons of real estate investors, capital would not continually flit from old to new assets. Instead, it would stay in place and mature as owners were forced to live with their decisions and consider the asset's future prospects. The likelihood of building to excess and "obsoleting" old buildings would decrease.

Transaction Taxes

When prices are bid up and returns are volatile, owners sell their holdings.[81] When assets are illiquid and transaction costs onerous, owners retain them. Through its fiscal policies, therefore, the public sector has a modicum of control over the exchange of real estate. Levying taxes, such as the capital gains tax, on the transfer of assets is one way in which the government affects the benefits that investors receive from a sale. The relatively low capital gains tax rates of the 2000s (reduced from 28 percent to 20 percent in 1981 and then again—to 15 percent—in 2003) presented no deterrent to transactions.[82] Similarly, the federal depreciation deduction reinforces short-termism and capital switching from older buildings to newer ones. Once building owners have taken their depreciation allowances, they seek new acquisitions through which they can reset the clock on these deductions.

The federal government could nudge capital gains and transaction taxes upward to act as a drag on exchange values, which would decrease asset flipping and price volatility.[83] Income from property ownership would need to be extracted more from operations than from sales, which would reinforce attention to current and future use values.

Similarly, reducing the value of depreciation deductions would decrease the after-tax returns from flipping property. Restructuring them so that they are bunched toward the end of average holding periods would encourage owners to retain their properties for a longer time. We could expect an effect opposite from that of the "Accelerated Cost Recovery System" instituted in 1981, which allowed commercial real estate investors to take these deductions early on in the building's life and led to the bubble and "see-through" building problem of late 1980s.[84]

Life Cycle and Reuse Planning

Local planning policies can also make developers more future-oriented. Most property owners already plan for the future on an ad hoc basis; for example, their building engineers consider the longevity of building systems to estimate when fixtures will have to be replaced. Formalizing future thinking through life cycle and reuse planning before a new building breaks ground could help stave off obsolescence.[85] Planning, Lynch reminds us, involves seeing "the spatio-temporal whole."[86]

If encouraged to think more about future values at the concept and design stages, developers might erect more adaptable and enduring structures.[87] As a guiding example, architect James Goettsch's Blue Cross–Blue Shield Tower in Chicago's East Loop submarket was built for vertical expansion. The building was constructed in two phases, with the second phase adding twenty-four stories on top of a fully operational thirty-three-floor skyscraper. Designers considered everything from how the exteriors of the two sections of the building—clad in glass, stainless steel, and stone—could be matched as it grew to how atrium spaces could be converted into high-zone elevators without interrupting work flows. Although nearly a decade separated the two phases of construction, no visible distinction between the old and new portions of the building is apparent.

If developers had to think like planners, they might choose to build more modularly and flexibly to reduce the cost of future adjustments to interior usage changes. They might not build out an entire site, leaving temporary common space that could be converted to new uses if growth warranted it.[88] They might select low-carbon technologies to accommodate the long life cycles of buildings and the impacts of climate change (what some call "future-proofing" energy design).[89] They might even plan for their buildings' demise, something that is of increasing concern as towers compete for new heights and landfills swell with construction and demolition debris. Keeping the door open for future "deconstruction" is one of the intentions of the

Chicago-based ReBuilding Exchange, which works with developers to ensure that building materials are removed from the job site intact and available for resale and reuse.[90] In order to ease future disassembly, architects can integrate materials with a high potential for recyclability, avoiding composite products and permanent glues, such as contact cement or epoxy, that make separation difficult.[91]

Less voluntary policies would compel developers to commit capital to future modernization or reuse. While property owners already set aside a portion of their operating income in reserves (for replacing heating systems), those reserves are often underfunded, and renovations are considered only when systems are on the brink of collapse. Developers could be required to buy property insurance or guarantee bonds on new commercial structures to guard against their obsolescence and abandonment, reducing the costs assumed by the public sector in dealing with junked properties.[92] These safeguards would ensure that at the end of a structure's useful life, the financial resources to fund the demolition or conversion of the structure would be available.[93]

Perpetual Surplus?

Any effort to stem overbuilding would be unpopular; the public sector would need to proceed cautiously and ensure that measures intended to address this problem were properly designed to match both the cadence of real estate cycles and the specificity of particular markets. It would also require some suspension of disbelief. One must imagine, for instance, responsive governments more interested in the collective welfare of residents than, in Leviathan-like fashion, amassing wealth in their own coffers. One must set aside the fact that many elected officials and administrators have worked for the real estate industry and that even more, regardless of their professional origins, play a role in the very overbuilding they are positioned to regulate. One must assume that governments possess the ability to correctly assess the real-time status of property markets and to schedule their interventions accordingly. While distinguishing a bubble, in which production and pricing bear little relation to occupant demand, from a spike due to growth or technological innovation (or a "rational" bubble caused by uncertainty or disagreement about asset values) may not be impossible, at the very least it will require experience-honed powers of judgment and the will to use them.[94] One also must trust governments to make decisions under great uncertainty and complex political pressures.

At the same time, recommendations for interventions to modulate the speed and amplitudes of real estate cycles are not radical propositions, given the strong interdependencies between collective action and property markets. Property markets have historically been "embedded" in formal regulations and "performed" by the practices and common understandings of professionals engaged in their constitution. As I have demonstrated, boom-bust cycles are not laws of nature; the financial liberalization and herding behaviors at the root of overbuilding episodes have public and institutional origins. Therefore, reconfiguring and reregulating property markets with countercyclical intent would be well within the historical purview of both the public sector and professional associations.

But would preventing overbuilding and encouraging building modernization justify the costs associated with the policies needed to accomplish these goals? What if the public sector mistakenly restricted construction in the absence of an actual speculative bubble? Would the costs of such interventions be higher than the costs of inaction when action was needed? To answer these questions, the net welfare effects of the three sets of policies introduced in this chapter would need to be determined both nationally and in specific regional contexts.

The most localized and targeted measures (such as requiring developers to submit reuse plans) would limit public costs, but would also be more easily undermined by loopholes and noncompliance. In contrast, measures with a broader reach (such as raising interest rates) would be more difficult to circumvent and hence potentially more effective. But these latter measures would also be more costly to implement and might have more serious side effects. For example, stricter monetary policy could potentially hurt sectors of the economy other than real estate that are not exhibiting signs of overheating. Reduced output, lower employment, and inflation could ensue.

One of the most serious effects of intentionally slowing construction in advance of a bubble would be the potential for regulation-induced appreciation. If development controls were imposed at a regional or national level, for example, occupants would be unable to move to escape their effects, and owners could raise prices. Although it depends on the specific growth control enacted and the type of market affected, there is a correlation between those regions with more restrictive land use regimes and those with higher property values.[95] If policies intended to stem overbuilding also increase exclusion (e.g., by decreasing the affordability of housing or commercial space), then they should be reexamined or designed to lessen such effects.

But before assuming a necessary trade-off between slower growth and higher prices, it is important to recognize that price increases are caused by both decreases in supply *and* increases in demand. If new supply is restricted while demand grows, property values increase, but residents and property owners may be ostensibly better off. For example, studies of whether Portland's growth boundaries raised prices found that the policy did have a noticeable appreciative effect, but that it was due more to increased housing demand, employment growth, and rising incomes than to restricted development opportunities.[96] In other words, restricting overbuilding may result in more pleasant places to live and work with improved public services, and this enhanced amenity value may be reflected in higher prices.

Actual demand and income growth, therefore, are key to understanding the extent to which policies such as growth controls or transaction taxes will make the available supply less affordable. In the absence of strong demand, restricting supply may not have any observable effect on prices. If new supply results from an influx of capital (and not from a demand surge), stemming that influx might even suppress prices, as it would bring them more into line with what occupants can actually pay.[97] Appreciation, therefore, would be less likely if regulations could be confined to those times and places prone to financially driven overbuilding.

Salvaging the Past, Constructing the Future

The goal is not to shut down the party and take the fun out of development. Nor is it to cling to the old remains of cities out of a misplaced sense of reverence, fear, or guilt.[98] No building will last forever, nor should it, and no consensus exists about how fast cities should change. In the words of architectural critic Rayner Banham, "We still have not formulated intellectual attitudes for living in a throwaway economy. We eagerly consume ephemera . . . but insist on aesthetic and moral standards hitched to permanency, durability and perennity."[99]

The suggestions above propose a middle ground between keeping and abandoning. They urge a shift from envisaging incessant expansion to managing the ongoing modification of inherited environments.

These interventions would have slowed the flow of new construction in Chicago during the 2000s. Developers would have built new only when the mandate to do so was less ambiguous—for example, a commission from an expanding tenant or one new to the region. The weight of new regulations would have suppressed tenants' replacement demand by making new construction a less cost-effective choice than growing or changing in place. With extended time horizons, landlords would have updated their older inventories rather than disinvesting in or replacing them when they appeared outdated. The ravages of obsolescence would have been staved off. Younger generations and businesses might have viewed building recycling as a sustainable antidote to overconsumption, and vintage buildings with modern interiors would have developed more cachet. A more regulated stream of better-occupied new construction would have allowed city planning to realize its potential, targeting services and infrastructure to those areas most in need.

With the right encouragement, refurbishment and structural modifications may yet become the dominant construction activities in the years ahead. If they do, it is possible that we will avoid the mistakes of the past while preparing for a future of constrained resources and financial volatility. After all, the goal should be to live in a deliberately modified world instead of a disposable one.

Epilogue

Why We Will Continue to Overbuild

Recession-Plagued Nation Demands New Bubble to Invest In
THE ONION HEADLINE, July 14, 2008

The cranes have returned to the Loop.

Most of the major office developers in Chicago emerged from the financial crisis scraped and scratched but substantially richer as a result of their participation in the Millennial Boom. By 2012 they were readying themselves for the next one. "In Chicago, we love new [office] buildings. Everyone is waiting for the next one . . . it's just a matter of time."[1]

Hines, for example, proposed a fifty-plus-story tower for the western bank of the Chicago River. When lenders turned down the developer's request for a construction loan and anchor tenants reneged on their letters of intent, the firm said it would finance the spec office project itself. Hines came back with a smaller skyscraper and the financial support of Canadian property firm Ivanhoe Cambridge. It hired the best brokers and filled almost 65 percent of the building with Chicago-based tenants, such as law firm McDermott Will & Emery. It received entitlements and promises of TIF assistance from the city. The project broke ground in 2013, leading observers to wonder: would Hines's River Point be the project, akin to John Buck's One Wacker Drive in the late 1990s, that kicks off the *next* downtown office boom?

Developers have submitted proposals to add more than 8 million square feet to the downtown Chicago office market.[2] About fifteen high-profile sites are being held with options to develop. Some are surface parking lots (such as 625 West Adams), others vacant land (such as 130 North Franklin, where the Chicago Mercantile Exchange once stood). All but a few of these sites are in the West Loop. Capital markets are back "with a vengeance" and looking for deals.[3] Another Hines trophy building, the sixty-story tower at 300 North LaSalle (completed in 2009), was sold in 2010 for a record price at a time when most thought the commercial property market had bottomed out after the recession. It then sold again four years later for another record price: $850 million (a 30 percent increase from 2010).[4]

The conversation has shifted from one of excess to one of scarcity. Ignoring the lingering vacancies in towers built during the almost-forgotten boom of the 1980s, the talk is of a shortage of well-located buildings with enough contiguous space to accommodate large tenants. Brokers use words like "bursting" and "pent-up" to describe the demand that has welled up during the five-year lull.[5] They point to the fact that several high-profile companies have relocated from suburban office parks to downtown locations and that average rents are inching upward.[6]

Downtown office vacancies have declined in many cities, including Chicago. In 2013 the city's published vacancy rate fell to under 15 percent from a high of 18 percent.[7] Although 15 percent is still a far cry from the 9.5 percent rates at the start of the Millennial Boom, it appears as though developers have become accustomed to building in overbuilt markets. What passes for an acceptable vacancy rate has been creeping upward. As one developer noted:

> The market is rebounding as we speak [in 2012]. We used to look at vacancy rates and say 10 percent was the threshold. If the market goes below 10 percent, you build. For this next cycle, I'm looking for 12 percent. You see, 12 percent is the new 10 percent . . . I think we're going to hit 12 percent in 2016.[8]

Even the hard-hit East Loop has experienced renewed interest, less as an office location than as a retail and residential one. In an article entitled "Michigan Avenue's Dead Zone Comes Back to Life," developers expressed intentions to erect four new hotels, two apartment towers, and new retail space there.[9] Meanwhile, residential developers never really stopped building, and office developers turned their sights to new apartments and condo towers. Almost 5,000 new apartment units were completed by the end of 2014, and another 3,000 are expected to be finished in 2015.[10] The pop-up shops and temporary art galleries that were useful during the bust as a way to animate empty space are being asked to kindly vacate the premises.

And yet absorption rates for apartments have averaged only 1,300 units a year, much lower than the 4,000 the market would need to keep pace. Only ten major companies have leases expiring between 2016 and 2019, and they may choose not to relocate.[11] Large tenants are reducing their footprints or subleasing space to other companies. Even law firm McDermott Will & Emery said it would take 10 percent less space than it previously occupied when it moves to River Point.[12]

Chicago is still doing better than many of the other cities that built to excess during the Millennial Boom. Downtown office markets in the Southwest and Southeast have not recovered; Miami, San Jose, Atlanta, San Diego, and Dallas had vacancy rates hovering at or above 20 percent in 2014. Nonetheless, banks and developers in those areas report that they too are reloading in anticipation of the next wave of new construction.[13]

REPEAT PERFORMANCE

Construction activity comes in spurts. Periods of above-average building activity alternate with times when new construction starts are atypically low. Cycles seem to follow a regular schedule—first, the upturn, then the downturn, and after a short pause, building picks up again. The rhythms of urban change appear synchronized, inevitable.

If, as in a cycle, the future is not so different from the past, then the future is reassuringly calculable and predictable. Even Homer Hoyt's intention in tracking Chicago land values almost a century ago was to predict the next boom and bust so that investors would not be wiped out in the future (Hoyt lost his shirt during the Great Depression). The cycle analogy orders time in a way that makes the future less threatening. Real estate actors like to insist that nothing is ever really new; whether activity is picking up or slowing down, they have "seen it before." The predictive certainty a cycle implies almost seems to bolster participants' sense of confidence and encourage their risk taking, making its ups and downs a self-fulfilling prophecy.

At the same time, Hoyt famously suggested that real estate cycles were just long enough (eighteen years in his mind) for the lessons of the previous cycle to be forgotten.[14] Indeed, for many in the industry, the commercial overbuilding of the Millennial Boom is already a distant memory.

Neither overconfidence nor ignorance, however, can entirely explain why we will continue to overbuild. While believing in the inevitable return of booms and forgetting about the associated dangers surely contribute to their reappearance, other conditions will be required to jump-start the next construction jag.

For one, new construction must be profitable. New construction represents the apotheosis for developers who are anxiously sitting on parcels of land and older buildings. If they can find an anchor tenant willing to leave its current premises and convince investors and banks to open their wallets, developers will break ground. And if they are successful, others will follow suit. A developer's success depends on the assistance of allied real estate professionals—brokers, attorneys, appraisers, local governments, architects, and moving companies—who work with tenants to co-construct the market for their products. Like developers, these professionals are also waiting for the signal, giddy with anticipation, as they too will make above-average profits during the boom-time rush.

Second, credit must be available to lubricate the machinery of development. Neither Dodd-Frank nor the individual settlements with the investment banks responsible for the worst misconduct during the Millennial Boom will temper the ingenuity of finance or the desire for additional liquidity in property markets. The sale of novel debt instruments will enroll investors and recapitalize the banks, which will originate more loans in response. If cities, equity investors, and developers can borrow to create new assets or reposition old ones, their indebtedness will produce income streams that can be converted into instruments traded by other investors. By taking as much of this "hot money" as they can grab, they will inflate valuations. Asset pricing will reflect the herky-jerky movements of volatile financial markets, more so than the incremental movements in local population and employment. Plans and proposals hatched during the quietude of the bust will be given material form when a surge of public and private capital switches into and inundates the real estate sector.

To keep pace with such expansions, tenant demand will need to grow—either because tenants are growing in size or taking more space per worker. And if demand does not increase, new buildings will either sit vacant or, more likely, be filled with occupants drawn from marginally older buildings. New buildings will make older ones "obsolete" as creative destruction justifies more new development, more public subsidies, and more financialization.

Given the speed at which capital circulates through these restless cities, occupying and quickly withdrawing from buildings, it may seem like a Promethean task to slow the pace of new construction and obsolescence. Breaking these chain reactions and reining in surpluses will require more than city planning; capital controls at the national and global scales, as well as new ways of thinking about both the future and the past, will be necessary to encourage better use and upkeep of the cities we have inherited and that we in turn will pass along.

Appendix

Interviews quoted (by chapter of first mention)

Chapter 1

Broker A, global brokerage firm, January 22, 2007 and March 28, 2007
Developer B, national commercial real estate firm, April 17, 2009
Tenant A, global law firm, October 16, 2006
Broker D, global brokerage firm, June 6, 2014
Developer F, residential development firm active in the Midwest, February 10, 2014
Facilities Manager A, global law firm, February 14, 2013

Chapter 2

Developer A, residential development firm active in the Midwest, October 16, 2012
Lender A, real estate and construction division, money center bank, February 10, 2010
Lender E, commercial real estate division, money center bank, May 3, 2012
Developer C, commercial development firm active in the Midwest, October 16, 2012

Chapter 3

Lender B, capital markets and real estate group, money center bank, June 4, 2012

Pension Fund Advisor A, global investment management firm, November 2, 2010

Broker G, local retail brokerage, December 8, 2006

Broker C, global real estate services firm, February 9, 2007

Investment Broker A, national real estate intermediary, March 4, 2011

Broker F, brokerage services division at local development firm, December 6, 2012

Real Estate Consultant A, boutique firm, October 16, 2012

Appraiser E, national real estate valuation and consulting firm, June 12, 2013

Appraiser A, national real estate valuation and consulting firm, December 11, 2012

Appraiser B, national real estate valuation and consulting firm, June 12, 2013

Architect A, Department of Buildings of City of Chicago June 15, 2006

Lender D, construction division, money center bank, February 10, 2010

Developer D, local commercial development firm, December 6, 2012

Planner B, Department of Housing and Economic Development of City of Chicago, August 3, 2009

Chapter 4

Broker B, boutique office brokerage firm, August 29, 2012

Chapter 5

Reporter B, local business newspaper, January 5, 2011

Mortgage Broker A, commercial real estate capital intermediary, February 17, 2012

Asset Manager A, national real estate investment trust, April 20, 2012

Mortgage Broker B, commercial real estate capital intermediary, April 8, 2012

Lender C, banking and real estate division, national law firm, March 17, 2010

Lender G, former employee of large midwestern bank, March 6, 2014

Planner G, Department of Housing and Economic Development of City of Chicago, March 1, 2012

Planner A, Department of Housing and Economic Development of City of Chicago, August 29, 2012

Chapter 6

Broker H, brokerage services division of national development firm, April 2, 2103

Planner C, Department of Housing and Economic Development of City of Chicago, February 5, 2014

Acknowledgments

I try to convey to my doctoral students the importance of researching a topic of significance in the field. The last question you want to hear after devoting nine of the best years of your life to one project is "So what?" A significant research project, I explain, examines a topical phenomenon, responds to an absence, or corrects a prior theoretical or empirical misunderstanding.

But, the clever students ask, doesn't it all boil down to marketing? After all, you can hype any mundane problem into a critical, do-or-die dilemma. Gaps in the literature, on their own, do not portend interesting research. Perhaps there is a good reason why no one has ever written a dissertation on the geography of pet-grooming services. Shouldn't you just be concerned with getting the job done—that is, with satisfying the standards of your specific professional community?

I still hold out hope for significance. If I were only interested in meeting the standards for what is considered a passable book in my field of urban planning, I would have written a very different one. I would have contributed, perhaps, another hard-hitting but dreary piece of critical political economy that bemoans the inevitable crisis of this or that. Or I would have written an uplifting, but under-theorized, text on the possibilities of progressive planning. Instead, I wrote the book I wanted to write and have taken some risks that may not have been advisable in retrospect. For example, my approach is robustly interdisciplinary, bringing together ideas and authors that rarely speak to each other. I have sought to create a readable blend of theory and empirics, academic and popular writing. I question the necessity of new construction, which, in some circles, is an invitation for serious condemnation.

The book is also personally significant. It focuses on my adopted home town, Chicago, which fascinates and frustrates me on a daily basis. And it is grounded in a deeply held belief in the value of junk. I come from a long line of garage-salers, used clothing purveyors, and—if you go back a century—rag-pickers, who taught me not only how to spot a bargain but also instilled a healthy suspicion about claims of "new-and-improved." (My sister, Lauren Weber, does a great job weaving our family's narrative into a rich social history of thrift in her book, *In Cheap We Trust: The Story of a Misunderstood American Virtue* [New York: Little Brown, 2009].) Possessing an unstudied appreciation for old buildings and neighborhoods that is part Jane Jacobs and part Stewart Brand, my gut instinct is to side with the second-hand and slightly shopworn. At the same time, I share with many of my fellow planners the hopeful, modernist conviction that certain design and technological changes can tangibly enrich the lived experience of city dwellers. The book attempts to balance these competing sentiments.

As is true of all forms of academic production, this book is the result of many years of collaboration. For their wise counsel, technical assistance, and good humor, I am indebted to my faculty and student colleagues in the College of Urban Planning and Public Affairs at the University of Illinois at Chicago: Phil Ashton, Mirabai Auer, Teresa Cordova, Marc Doussard, Glenda Garelli, Robby Habans, Charlie Hoch, Nancy Hudspeth, April Jackson, Jihwan Kim, Dave Merriman, Mike Pagano, David Perry, Brent Ryan, Lucinda Scharbach, Janet Smith, Ben Teresa, Nik Theodore, Aaron Werner, Curt Winkle, and Moira Zellner. Other colleagues at UIC and elsewhere were also exceedingly generous with their advice and insight, including Manuel Aalbers, Bob Beauregard, Larry Bennett, Bob Bruegmann, Scott Campbell, Gordon Clark, Win Curran, Judith De Jong, Susan Fainstein, Stephanie Farmer, Jean Guarino, Sharon Haar, Brad Hunt, Dennis Judd, Dan McMillen, Faranak Miraftab, Tony Orum, Andy Pike, Susanne Schnell, David Wilson, and Peter Wissoker. The Great Cities Institute provided me with a Faculty Scholarship and, along with the Urban Land Institute, supported the Nicholas Trkla Award, a scholarship that funded four amazing research assistants: Joel Benedetti, Michael Kaplan, Ian Ludwig, and Curtis Witek. Robby Habans deserves special mention for punching up the manuscript graphically. The Narrative Non-Fiction Group at the Institute for Research on Race and Public Policy—Michelle Boyd, Anna Guevarra, Brenda Parker, Karen Su—instilled in me a deep respect not only for narrative arcs and vivid prose, but also for writing support groups.

I participate in several groups—including the Chicago Realty Club and the Richard T. Ely Chapter of Lambda Alpha International—whose fellow members became interview subjects and friends. There is no more savvy and well-connected an association than my Sunday morning book club—Martha

Frish, Linda Goodman, Fran Grossman, Mary Ludgin, Roberta Nechin, Kathy Sandstrom, and Christine Williams—who have provided me with the boldfaced names and backstories for every real estate project in town.

At the University of Chicago Press, I am indebted to Robert Devens for believing in the intellectual merit of studying real estate and to Tim Mennel for his subject matter expertise, editorial knife skills, and hand-holding. The reviews my editors solicited vastly improved the final product. Thank you also to Nora Devlin, Mark Reschke, Laura Bevir, Norma Sims Roche, and Ashley Pierce.

Writing a book requires the kind of *sitzfleisch* that is difficult to come by in this era of constant distraction. I would have been unable to concentrate on this topic for days on end were it not for my friends and relations, who prodded me with questions, edited early versions of chapters, and offered me a quiet place to write when home and work proved too engrossing. Thank you: Phil Ashton, Jon DeVries, Mark Donovan, Will Elison, Fran Grossman, Ted Grossman, Sara Hymowitz, Sharon Jacobs, Jim Loellbach, Dave McBride, Marc Neves, Lisa Perez, Sara Slawnik, Jacob Slichter, Janet Smith, Josh Steinhauer, Sarah Steinhauer, Frieda Stolzberg, Estelle Tarica, Bonnie Tawse, Evalyn Tennant, Ben Weber, Corinne Weber, David Weber, Donald Weber, Lauren Weber, Christine Williams, Merrill Zack, and Elise Zelechowski.

The book was a family affair—my journalist sister commiserated; my father and mother copyedited; and my brother provided the soundtrack for late-night writing sprees. My children wore me out and replenished me. My stature in their eyes dropped a few degrees when I let them know there would be no movie or made-for-TV anything based on this book. The primary inspiration for this project was my husband, John Slocum, whose creativity is matched only by his love and encouragement.

Notes

Introduction

1. *Life*, October 1986, 21–23, 27.

2. City of Chicago, *LaSalle Central Redevelopment Plan* (Chicago: Department of Housing and Economic Development, 2006), https://www.cityofchicago.org/dam/city/depts/dcd/tif/plans/T_147_LaSalleCentralRDP.pdf.

3. Joseph Schumpeter, *Capitalism, Socialism and Democracy* (New York: Harper, 1975).

4. Ibid., 87.

5. John P. Koval, Larry Bennett, Michael I. J. Bennett, Fassil Demissie, Roberta Garner, and Kiljoong Kim, eds., *The New Chicago* (Philadelphia: Temple University Press, 2006).

6. Homer Hoyt, *One Hundred Years of Land Values in Chicago: The Relationship of the Growth of Chicago to the Rise of Its Land Values, 1830–1933* (Chicago: University of Chicago Press, 1933), 260.

7. Carl Smith, *The Plan of Chicago: Daniel Burnham and the Remaking of the American City* (Chicago: University of Chicago Press, 2006).

8. John Hudson, *Chicago: A Geography of the City and Its Region* (Chicago: University of Chicago Press, 2006), 15.

9. In common parlance, "the Loop" often refers to the city's entire central business district, but formally, it is only the area surrounded on all sides by the downtown routes of the elevated train tracks. For this reason, I generally use the term "downtown" to describe the city's central business district and "the Loop" to refer to its core. As I explain in chapter 4, I adopt roughly the same extended geography that the City of Chicago uses to define the downtown for the purposes of collecting and analyzing property market statistics.

10. In order to encourage candor and to meet the standards for the University of Illinois at Chicago Institutional Review Board, I assured subjects that their identities would be kept confidential. Throughout the book, quotes and paraphrased information from the interviews are attributed using identifiers such as "Developer B" or "Broker E." I include an appendix of interviews and the dates on which they took place.

11. A "field" is a kind of structured social space in which agents act out consistent roles. Pierre Bourdieu, *The Field of Cultural Production* (New York: Columbia University Press, 1993).

12. Mass securitization has exposed houses and households to similar pressures, but owner-occupied, single-family homes have a distinct capital structure, their smaller scale affords own-

ers more control over the pace and nature of change, redevelopment of residential neighborhoods is more often contested by community groups, and change is differently motivated.

13. The federal government does not collect data on commercial real estate, a void subsequently filled by the handful of private subscription services that track new construction as
well as lease and sale transactions across the country. I rely almost exclusively on data collected
by CoStar, a private provider of commercial real estate information that was founded in 1987.
CoStar maintains a team of almost 1,000 researchers who gather primary information through
direct counts and site inspections, as well as through document reviews, news monitoring, and
secondary data analysis. See http://www.costargroup.com/.

14. M. R. G. Conzen, *Thinking about Urban Form: Papers on Urban Morphology, 1932–1998*
(Oxford: Peter Lang, 2004). See also J. W. R. Whitehand, "Development cycles and urban landscapes," *Geography* 79, no. 1 (1984): 3–17.

15. With its focus on these relatively novel actors and instruments, this book extends Susan Fainstein's landmark study of urban redevelopment and Karl Beitel's prescient analysis of
finance-driven booms into the twenty-first century. See Fainstein, *The City Builders: Property,
Politics, and Planning in London and New York, 1980–2000*, 2nd ed. (Lawrence: University Press
of Kansas, 2001); and Beitel, "Financial cycles and building booms: A supply side account,"
Environment and Planning A 32 (2000): 2113–32.

16. U.S. Bureau of Economic Analysis, *Value Added by Industry as a Percentage of Gross Domestic Product*, 2009, http://www.bea.gov/newsreleases/industry/gdpindustry/gdpindnewsrelease
.htm. These figures do not include construction or finance and insurance, which contribute an
additional 4.1 percent and 7.5 percent, respectively, to total output.

Chapter One

1. Justin Davidson, "The glass stampede," *New York Magazine*, September 7, 2008.

2. Atlanta Regional Commission, "Monthly Building Permits, Comparison with Top-15
Metros," *The Quarter* 3, 2010, http://documents.atlantaregional.com/enewsletters/theQtr/the
-quarter.html.

3. Rem Koolhaas, *Delirious New York* (New York: Monacelli Press, 1997).

4. Henry James, *The American Scene* (London: Chapman & Hall, 1907), 232.

5. Marshall Berman, *All That Is Solid Melts into Air* (New York: Penguin, 1982), 15.

6. Zygmunt Bauman, *Liquid Modernity* (New York: Polity Press, 2000). See also Max Page,
The Creative Destruction of Manhattan, 1900–1940 (Chicago: University of Chicago Press, 1999).
Page shows how New Yorkers were taught to glorify the notion of a constantly changing envi-
ronment and to think of it as progressive during a period of major building and demolition.

7. John Jakle and David Wilson, *Derelict Landscapes: The Wasting of America's Built Environment* (Savage, MD: Rowman & Littlefield, 1992), 284.

8. Anne Vernez Moudon, *Built for Change* (Cambridge, MA: MIT Press, 1986).

9. David Harvey, *The Condition of Postmodernity* (Malden, MA: Blackwell, 1990).

10. Anique Hommels, *Unbuilding Cities: Obduracy in Urban Socio-technical Change* (Cambridge, MA: MIT Press, 2005), 10.

11. Kevin Lynch, *What Time Is This Place?* (Cambridge, MA: MIT Press, 1976, 2001), 38.

12. Paul Pierson, *Politics in Time: History, Institutions, and Social Analysis* (Princeton, NJ:
Princeton University Press, 2004).

13. Lynch, *What Time Is This Place?*, 99.

14. Paul Knox goes so far as to label entire cities either "slow" or "fast" based on their average pace of change. "Creating ordinary places: Slow cities in a fast world," *Journal of Urban Design* 10, no. 1 (2005): 1–11.

15. Aldo Rossi, *The Architecture of the City* (Cambridge, MA: MIT Press, 1982), 22.

16. For a review of cycle theory in economics, see Stephen Pyhrr, Stephen Roulac, and Waldo Born, "Real estate cycles and their strategic implications for investors and portfolio managers in the global economy," *Journal of Real Estate Research* 18, no. 1 (1999): 7–68; William Wheaton, "Real estate 'cycles': Some fundamentals," *Real Estate Economics* 27 (1999): 209–30; Glenn Mueller, "Understanding real estate's physical and financial market cycles," *Real Estate Finance* 12 (1995): 47–52; Glenn Mueller, "Real estate rental growth rates at different points in the physical market cycle," *Journal of Real Estate Research* 18, no. 1 (1999): 131–50.

17. Simon Kuznets, *Secular Movements in Production and Prices* (Boston: Houghton Mifflin, 1930).

18. Wesley Clair Mitchell, *Business Cycles: The Problem and Its Setting*, Publications of the National Bureau of Economic Research, no. 10 (New York: National Bureau of Economic Research, 1927), www.nber.org/chapters/c0680; Homer Hoyt, *One Hundred Years of Land Values in Chicago: The Relationship of the Growth of Chicago to the Rise of Its Land Values, 1830–1933* (Chicago: University of Chicago Press, 1933).

19. Eugene Fama, "Random walks in stock market prices," *Financial Analysts Journal* 21 (1965): 55–59.

20. See, for example, William Wheaton, "The cyclic behavior of the national office market," *Real Estate Economics* 15, no. 4 (1987): 281–99; Helga Leitner, "Capital markets, the development industry, and urban office market dynamics: Rethinking building cycles," *Environment and Planning A* 26, no. 5 (1994): 779–802; John Hekman, "Rental price adjustment and investment in the office market," *Real Estate Economics* 13, no. 1 (1985): 32–47. Convergence is likely due to the interpenetration of markets by national firms, contagion effects, and, as I will suggest, collective assumptions of how fast buildings age and need to be replaced.

21. Leo Grebler and Leland Burns, "Construction cycles in the United States since World War II," *Real Estate Economics* 10 (1982): 123–51, 125.

22. Lynch, *What Time Is This Place?*, 65.

23. Philip Mirowski, *More Heat than Light: Economics as Social Physics, Physics as Nature's Economics* (Cambridge: Cambridge University Press, 1991); Marieke De Goede, *Virtue, Fortune and Faith: A Genealogy of Finance* (Minneapolis: University of Minnesota Press, 2005).

24. Berman, *All That Is Solid Melts into Air.*

25. Lynch, *What Time Is This Place?*, 63.

26. Henri Lefebvre notes that while urban change follows a rhythm, "there is always something new and unforeseen that introduces itself into the repetitive: difference." *Rhythmanalysis: Space, Time, and Everyday Life*, trans. and ed. Stuart Elden (New York: Continuum, 2004), 6.

27. Michael Ball suggests we look at cycles as if analyzing the rings of a tree trunk: the age share of different building stocks varies in size depending on whether buildings in each "ring" were constructed during high-growth years (thick) or low-growth ones (thin). "Is there an office replacement cycle?," *Journal of Property Research* 20, no. 2 (2003): 173–89.

28. William Goetzmann and Frank Newman, "Securitization in the 1920's" (working paper 15650, National Bureau of Economic Research, 2010), http://www.nber.org/papers/w15650.pdf.

29. David Birch, Susan M. Jain, William Parsons, and Zhu Xiao Di, *America's Future Office Space Needs* (Washington, DC: National Association of Industrial and Office Parks, 1990).

30. See Table 1.1.

31. In Manhattan more office towers were built during the 1980s cycle to accommodate a growing service industry, but the Millennial Boom saw the addition of significantly more residential square footage as many of these office-using industries migrated to other parts of the New York region and the demand for apartments increased. Robert Beauregard, "The textures of property markets: Downtown housing and office conversions in New York City," *Urban Studies* 42, no. 13 (2005): 2431–45.

32. Some scholars have used the concept of a "replacement rate" (i.e., the fraction of the existing stock in some baseline year that was demolished and rebuilt within a particular period of time) to capture the spatial passage of time and changing face of the city. See Ball, "Is there an office replacement cycle?"; J. W. R. Whitehand, "Long-term changes in the form of the city centre: The case of redevelopment," *Geografiska Annaler* (1978): 79–96; Richard Dye and Daniel P. McMillen, "Teardowns and land values in the Chicago Metropolitan Area," *Journal of Urban Economics* 61 (2007): 45–63.

33. Davidson, "Glass stampede."

34. Southern California Association of Governments, "SCAG Region Demolition Permit Statistics: 1/1/2001 to 1/1/2011," *Regional Housing Needs Assessment,* 2011, http://www.scag.ca .gov/Documents/RHNAFinalMethodologyAppendices110311.pdf.

35. John Logan and Harvey Molotch, *Urban Fortunes: The Political Economy of Place* (Berkeley: University of California Press, 1987); Neil Smith, "Gentrification and uneven development," *Economic Geography* 58 (1982): 139–55.

36. Logan and Molotch, *Urban Fortunes.*

37. Textbooks on real estate economics typically adopt this perspective. See, for example, Richard Muth, *Urban Economic Problems* (New York: Harper & Row, 1975); Ed Mills, ed., *Dynamic Urban Models: Handbook of Regional and Urban Economics* (Amsterdam: Elsevier, 1987). Some economic models, however, more explicitly acknowledge that construction results from interactions between the markets for space and the markets for assets. See, for example, Denise DiPasquale and William Wheaton, *Urban Economics and Real Estate Markets* (Englewood Cliffs, NJ: Prentice Hall, 1996).

38. Colin Lizieri, "Occupier requirements in commercial real estate markets," *Urban Studies* 40, no. 5–6 (2003): 1151–69.

39. Colin Lizieri, *Towers of Capital: Office Markets and International Financial Services* (Oxford: Wiley-Blackwell, 2009).

40. It is "elastic"; as rents rise, building values follow a similar upward trajectory and additional supply is produced. See Lizieri, *Towers of Capital,* for a review, as well as Richard Voith and Theodore Crone, "National vacancy rates and the persistence of shocks in U.S. office markets," *Real Estate Economics* 16, no. 4 (Winter 1988): 437–58.

41. Sharon Zukin, *Landscapes of Power: From Detroit to Disney World* (Berkeley: University of California Press, 1991); Neil Smith, *The New Urban Frontier: Gentrification and the Revanchist City* (New York: Routledge, 1996). However, some scholars writing in this vein leave open the possibility that the needs of individual tenants and society at large can be met through investment in real estate, particularly if the state is involved and committed to participatory planning and equitable outcomes. For example, in Susan Fainstein's treatment of the development industry in London and New York City in the 1980s, developers sought to make money for themselves and their financiers, but they did so by satisfying the needs of occupants and quotas for affordable housing, public space, and job training imposed by government planners. Fainstein,

The City Builders: Property, Politics, and Planning in London and New York, 1980–2000, 2nd ed. (Lawrence: University Press of Kansas, 2001). See also Susan Fainstein, *The Just City* (Ithaca, NY: Cornell University Press, 2010).

42. Logan and Molotch, *Urban Fortunes*.

43. Ibid., 26.

44. David Harvey, *The Urban Experience* (Baltimore: Johns Hopkins University, 1989), 190; see also his *The Limits to Capital* (Chicago: University of Chicago Press, 1982) and his *The Urbanization of Capital: Studies in the History and Theory of Capitalist Urbanization* (Baltimore: Johns Hopkins University Press, 1985). See also Paul Virillio, *Speed and Politics* (New York: Semiotexte, 1986).

45. See, for example Harvey, *Limits to Capital*; Logan and Molotch, *Urban Fortunes*; Michael Storper and Richard Walker, *The Capitalist Imperative* (Oxford: Blackwell, 1989); Robert Beauregard, *Voices of Decline: The Postwar Fate of US Cities* (New York: Wiley-Blackwell, 1994); and Fainstein, *City Builders*. These authors share Marxist roots, a suspicion of positivist methodologies, a concern for social justice, and an acknowledgment of the importance of a scholar's ideological standpoint.

46. It is important to recognize these as metaphors and not as empirical descriptions. Indeed, different naturalistic metaphors have been deployed throughout history to comprehend and diagnose abundance, excess, and waste in markets. Some reference air while others evoke water in some form (waves, liquidity, froth, floods). See Marieke De Goede, "Finance and the excess," *Zeitschrift für Internationale Beziehungen* 16, no. 2 (2009): 299–310.

47. Different property types are differently aligned with cycles; office and retail have been shown to be less synchronized with the general business cycle while residential and industrial are more so. Wheaton, "Real estate 'cycles'"; William Wheaton and Raymond Torto, "An investment model of the demand and supply for industrial real estate," *Real Estate Economics* 18, no. 4 (1990): 530–47; John Kling and Thomas E. McCue, "Stylized facts about industrial property construction," *Journal of Real Estate Research* 6, no. 3 (1991): 293–304.

48. Broker A; see also Joe Feagin, "The secondary circuit of capital: Office construction in Houston, Texas," *International Journal of Urban and Regional Research* 11, no. 2 (1987): 172–92.

49. Some dubbed this "overbuilding porn." See, for example, "Ruins of the second Gilded Age," *New York Times Magazine*, July 8, 2009.

50. Robin Banerji and Patrick Jackson, "China's ghost towns and phantom malls," *BBC News Magazine*, August 13, 2012, http://www.bbc.com/news/magazine-19049254.

51. Several empirical studies have found metropolitan variation in natural vacancy rates to be unusually large. See Voith and Crone, "National vacancy rates"; William Wheaton and Raymond Torto, "Vacancy rates and the future of office rents," *Real Estate Economics* 16, no. 4 (1988): 430–36; Stuart Gabriel and Frank Nothaft, "Rental housing markets, the incidence and duration of vacancy, and the natural vacancy rate," *Journal of Urban Economics* 49, no. 1 (2001): 121–49; James Shilling, C. F. Sirmans, and John Corgel, "Price adjustment process for rental office space," *Journal of Urban Economics* 22, no. 1 (1987): 90–100.

52. In one of the few studies to measure commercial overbuilding, the authors' unit of analysis is an individual quarter or periods of quarters in which a metropolitan area's growth in office inventory is one standard deviation or more above the long-term average growth in office inventory over the past twenty years. Such an approach does not allow for temporal aberrations within what is a rather long (twenty years) characterization of a cycle. Mark Gallagher and An-

thony Wood, "Fear of overbuilding in the office sector: How real is the risk and can we predict it?," *Journal of Real Estate Research* 17, no. 1 (1999): 3–32.

53. See, for example, William Wheaton, "The cyclic behavior of the national office market," *Real Estate Economics* 15: 281–99; and Edward Glaeser, Joseph Gyourko, and Albert Saiz, "Housing supply and housing bubbles," *Journal of Urban Economics* 64, no. 2 (2008): 198–217.

54. Commercial real estate trends tend to lag the overall economy by anywhere from 6 to 24 months. Richard Barras and D. Ferguson, "Dynamic modeling of the building cycle: Empirical results," *Environment and Planning A* 19 (1987): 493–520.

55. Ko Wang and Yuqing Zhou, "Overbuilding: A game theoretic approach," *Real Estate Economics* 28, no. 3 (2000): 493–522; Steven Grenadier, "The strategic exercise of options: Development cascades and overbuilding in real estate markets," *Journal of Finance* 51 (1996): 1653–79.

56. Avner Bar-Ilan and William Strange, "Investment lags," *American Economic Review* (1996): 610–22, 620.

57. Anthony Downs, *The Revolution in Real Estate Finance* (Washington, DC: Brookings Institution, 1985); Kerry Vandell, "Tax structure and natural vacancy rates in the commercial real estate market," *Real Estate Economics* 31, no. 2 (2003): 245–67; C. Alan Garner, "Is commercial real estate reliving the 1980s and early 1990s?," *Economic Review* 93, no. 3 (2008): 89–115.

58. John Kling and Thomas McCue, "Office building investment and the macroeconomy: Empirical evidence, 1973–1985," *Real Estate Economics* 15, no. 3 (1987): 234–55.

59. Harvey, *Urban Experience*.

60. For Harvey, overaccumulation is the production of surplus value caused by simultaneous growth in profits and scarcity of profitable investment outlets. Harvey, *Limits to Capital*. See also Robert Brenner, *The Boom and the Bubble: The US in the World Economy* (London: Verso, 2002); Giovanni Arrighi, *The Long Twentieth Century: Money, Power, and the Origins of Our Times* (London: Verso, 1994).

61. Keynes made a distinction between the normal workings of capitalism, which he called "enterprise," and the act of trying to anticipate and outwit short-term price movements, which he called "speculation." John Maynard Keynes, *The General Theory of Employment, Interest and Money* (London: Macmillan, 1936). Some orthodox economists also use the term but reserve its use for those investments that, while still rational, have a short time frame, are "overly" optimistic, or lack important information. See, for example, Stephen Malpezzi and Susan Wachter, "The role of speculation in real estate cycles," *Journal of Real Estate Literature* 13, no. 2 (2005): 141–64.

62. See Charles Kindleberger, *Manias, Panics and Crashes: A History of Financial Crises* (New York: Basic Books, 1978); Mitchel Abolafia and Martin Kilduff, "Enacting market crisis: The social construction of a speculative bubble," *Administrative Science Quarterly* 33, no. 2 (1988): 177–93; and Robert Shiller, *Irrational Exuberance* (New York: Random House, 2005), on the psychology of investment markets that results in bubbles.

63. The inflection points between booms and bubbles are difficult to pin down in both practice and theory. Theoretically, all financial expansions are bubbles if taking risks and making speculative investments are considered healthy features of capitalism. Practically, deviations from some kind of a norm or benchmark (e.g., average occupancy rates) are visible mostly in retrospect. Janet Roitman, *Anti-crisis* (Durham, NC: Duke University Press, 2013); Christian Marazzi, *Capital and Language: From the New Economy to the War Economy* (Los Angeles: Semiotext, 2008).

64. Institutionalism in economics and sociology stresses the notion that all economic ac-

tivity is embedded in social relations. Creating markets for, say, land has required historically specific institutions (e.g., regulations, professional practices) to partially separate it from these relations and convert it into a more liquid commodity. See Karl Polanyi, *The Great Transformation: The Political and Economic Origins of Our Time* (New York: Beacon Press, 1944).

65. For exceptions, see Fainstein, *City Builders*; Beauregard, *Voices of Decline*; Logan and Molotch, *Urban Fortunes*; Joe Feagin and Robert Parker, *Building American Cities: The Urban Real Estate Game* (Washington, DC: Beard Books, 2002); Martin Murray, *Taming the Disorderly City: The Spatial Landscape of Johannesburg after Apartheid* (Ithaca, NY: Cornell University Press, 2008); Mark Pendras, "Urban politics and the production of capital mobility in the United States," *Environment and Planning A* 41, no. 7 (2009): 1691–1706.

66. For example, the architect Rem Koolhaas, commenting on the fast pace of uncoordinated change, notes, "Not only does *the system itself now decide what is needed*, but today the need is met so quickly that its fulfillment can take place before any other part of the system has been apprised of the demand or allowed to absorb, register, or modify itself in accordance with it" [italics mine]. Koolhaas, *Project on the City* (Miami, FL: Taschen, 2001), 525. Benjamin Lee and Edward LiPuma argue that "capital creates a social totality that is in constant motion, that constantly destroys itself in creating and expanding itself." Lee and LiPuma, "Cultures of circulation: The imaginations of modernity," *Public Culture* 14, no. 1 (2002): 191–213, 203.

67. Quoted in Leitner, "Capital markets," 784.

68. Thomas Gieryn, "What buildings do," *Theory and Society* 31 (2002): 35–74, 37.

69. Susan Smith, Moira Munro, and Hazel Christie, "Performing (housing) markets," *Urban Studies* 43, no. 1 (2006): 81–98, 82.

70. See, for example, Gwenda Blair, *The Trumps: Three Generations That Built an Empire* (New York: Simon and Schuster, 2001); Ben Johnson, *Money Talks, Bullsh*t Walks: Inside the Contrarian Mind of Billionaire Mogul Sam Zell* (New York: Penguin, 2009); Joshua Olsen, *Better Places, Better Lives: A Biography of James Rouse* (Washington, DC: Urban Land Institute, 2003); Philip Klutznick and Sidney Hyman, *Angles of Vision: A Memoir of My Lives* (Chicago: Ivan R. Dee, 1991); and Tom Shachtman, *Skyscraper Dreams: The Great Real Estate Dynasties of New York* (New York: Little, Brown, 1991).

71. Andrew Ross Sorkin, *Too Big to Fail: The Inside Story of How Wall Street and Washington Fought to Save the Financial System—and Themselves* (New York: Penguin, 2010); Michael Lewis, *The Big Short: Inside the Doomsday Machine* (New York: W. W. Norton, 2011).

72. Erecting a skyscraper in Chicago during the Millennial Boom, for example, required a development staff of at least ten plus more than fifty different vendors and suppliers of services (not including the multiple subcontractors hired by the general contractor). Developer B.

73. See, for example, Tom Angotti and Peter Marcuse, *New York for Sale: Community Planning Confronts Global Real Estate* (Cambridge, MA: MIT Press, 2011); Feagin and Parker, *Building American Cities*.

74. Based on ethnographic methodologies and direct observation, these accounts demonstrate that market information is socially embedded and often ambiguous and show how it is interpreted, weighted, and skewed through specific professional behaviors. See Karen Ho, *Liquidated: An Ethnography of Wall Street* (Durham, NC: Duke University Press, 2009); Caitlin Zaloom, "Ambiguous numbers: Trading technologies and interpretation in financial markets," *American Ethnologist* 30, no. 2 (2003): 258–72; Donald MacKenzie and Yuval Millo, "Constructing a market, performing theory: The historical sociology of a financial derivatives exchange," *American Journal of Sociology* 109, no. 1 (2003): 107–45; Karin Knorr-Cetina and Alex Preda,

eds., *The Sociology of Financial Markets* (New York: Oxford University Press, 2005); Mitchell Abolafia, *Making Markets: Opportunism and Restraint on Wall Street* (Cambridge, MA: Harvard University Press, 2001); Harvey Molotch, "Place in product," *International Journal of Urban and Regional Research* 26, no. 4 (2002): 665–88; Michel Callon, "The embeddedness of economic markets in economics," in *The Laws of the Markets*, ed. Michel Callon (Oxford: Wiley-Blackwell, 1998), 1–57; Jeffrey Chwieroth, *Capital Ideas: The IMF and the Rise of Financial Liberalization* (Princeton, NJ: Princeton University Press, 2010); Michel Callon and Fabian Muniesa, "Economic markets as calculative collective devices," *Organization Studies* 26 (2005): 1129–50.

75. Andrew Abbott, *The System of Professions: An Essay on the Division of Expert Labor* (Chicago: University of Chicago Press, 1988), 4.

76. Some consider this industry to be a "class" in its own right. Harvey, for example, argues that all property owners occupy a similar class position in capitalist economies, possess the monopoly power accorded by legal ownership, and behave in individual ways that benefit one another as a collective. David Harvey, "Class-monopoly rent, finance capital and the urban revolution," *Regional Studies* 8, no. 3–4 (1974): 239–55.

77. U.S. Bureau of Labor Statistics, "Industries at a Glance: NAICS 531," 2008, http://data.bls.gov/timeseries/CES5553000001?data_tool=XGtable.

78. These agents create consensus about the qualities of buildings and the meanings of these qualities. "How could we describe, in practice and theory, the structures of competition within markets . . . if relations of similitude and dissimilitude between the goods that circulate could not be established?" Michel Callon, Cécile Méadel, and Vololona Rabeharisoa, "The economy of qualities," *Economy and Society* 31, no. 2 (2002): 194–217, 196.

79. Marc Weiss, *The Rise of the Community Builders: The American Real Estate Industry and Urban Land Planning* (New York: Columbia University Press, 2002); Jeffrey Hornstein, *A Nation of Realtors: A Cultural History of the Twentieth-Century American Middle Class* (Durham, NC: Duke University Press, 2005).

80. Michael Pomerleano, "Back to the basics: Critical financial sector professions required in the aftermath of an asset bubble," *Appraisal Journal* (April 2002): 173–81.

81. Peter Haas, "Epistemic communities and international policy coordination," *International Organization* 46, no. 1 (1992): 1–35.

82. Robert Shiller, "Historic turning points in real estate," *Eastern Economic Journal* 34, no. 1 (2008): 1–13; Jose Scheinkman and Wei Xiong, "Overconfidence and speculative bubbles," *Journal of Political Economy* 111, no. 6 (2003): 1183–1220.

83. "All human products have an irresistible tendency to become old, but the meaning of ageing is different for different objects and circumstances." Manfredi Nicoletti, "Obsolescence," *Architectural Review* 143, no. 6 (1968): 413–15.

84. Daniel Abramson, "Obsolescence: Notes toward a history," *Praxis: Journal of Writing Building* 5 (2003): 106–112, 110; see also John Bryson, "Obsolescence and the process of creative reconstruction," *Urban Studies* 34, no. 9 (1997): 1439–58.

85. Joseph Schumpeter, *Capitalism, Socialism and Democracy* (New York: Harper, 1975). See also George Nelson, "Obsolescence," *Perspecta* 11 (1967): 170–76. For a historical overview of the concept of progress and its use in urban planning, see Scott Campbell, "Is 'progress' no longer progressive? Reclaiming the ideology of progress in planning" (working paper 06-01, Urban and Regional Research Collaborative, University of Michigan, 2006).

86. Facilities Manager A and Tenant A. Looking at time-series data on office holdings in the United Kingdom over the second half of the twentieth century, Ball finds little correlation

between a building's age cohort and systematic depreciation. Ball, "Is there an office replacement cycle?"

87. Developer F.

88. Tenants receive an "allowance" from the developer to pay for the improvements above the most basic ones (called the "turnover condition"). The costs of the build-out beyond the allowance are the tenant's responsibility, as are the furniture, fixtures, and equipment (FFE).

89. In explaining similarities in behavior across professional fields, Paul DiMaggio and Walter Powell note the three mechanisms through which organizational homogeneity ("institutional isomorphism") occurs. Using their categories, I would argue that the desires of tenants mirror those of the suppliers of space not through *coercive* isomorphism, but rather through *mimetic* (modeling their preferences on those of others in the field to reduce uncertainty) and *normative* (engaging in similar forms of professional socialization) processes. See "The iron cage revisited: Institutional isomorphism and collective rationality in organizational fields," *American Sociological Review* 48 (1983): 147–60.

90. Amanda Robert, "Lawyers and real estate experts talk office space," *Chicago Lawyer*, March 4, 2011.

91. Broker D.

92. Ibid. The percentage of an office building's area shared by all tenants is called a "loss factor." It is the difference between the net (usable) and gross (billable) area.

93. Facilities Manager A.

94. This concept is commonly used in the field of science and technology studies. For a study of the design of computer software, see Steve Woolgar, "Configuring the user: The case of usability trials," in *A Sociology of Monsters: Essays on Power, Technology and Domination*, ed. John Law, Sociological Review Monograph 58 (London: Routledge, 1991), 66–75. See also Rachel Weber, "Manufacturing gender in commercial and military cockpit design," *Science, Technology & Human Values* 22 (1997): 235–53, for a study of aircraft design. For more general explanations of the concept, see Nelly Oudshoorn and Trevor Pinch, *How Users Matter: The Co-construction of Users and Technology* (Cambridge, MA: MIT Press, 2003).

95. The inability to disentangle demand from supply-side causes of construction brings to mind Alfred Marshall's famous scissor analogy: "We might as reasonably dispute whether it is the upper or the under blade of a pair of scissors that cuts a piece of paper, as whether value is governed by utility or cost of production." *Principles of Economics*, 8th ed. (London: Macmillan, 1920), 348.

Chapter Two

1. Carrie Jones, "Apartment house bonds: Some plans for reorganizing defaulted issues," *Journal of Land & Public Utility Economics* 10, no. 1 (February 1934): 67–77, cited in William Goetzmann and Frank Newmann, "Securitization in the 1920's" (working paper 15650, National Bureau of Economic Research, 2010), http://www.nber.org/papers/w15650.pdf.

2. For a discussion of financialization, see Robert Boyer, "Is a finance-led growth regime a viable alternative to Fordism? A preliminary analysis," *Economy and Society* 29 (2000): 111–45; Gerard Duménil and Dominique Lévy, *Capital Resurgent: Roots of the Neoliberal Revolution* (Cambridge, MA: Harvard University Press, 2004); Greta Krippner, "The financialization of the American economy," *Socio-Economic Review* 3 (2005): 173–208; and Giovanni Arrighi, *The Long Twentieth Century: Money, Power, and The Origins of Our Times* (London: Verso, 1994). See

also critiques by Brett Christophers, who questions whether sufficient evidence exists to support claims that such a "profound structural shift" has occurred: Christophers, "Revisiting the urbanization of capital," *Annals of the Association of American Geographers* 101 (2011): 1.

3. Phillip Cerny, "The political economy of international finance," in *Finance and World Politics* (New York: Edward Elgar, 1993), 18.

4. James Crotty, "Structural causes of the global financial crisis: A critical assessment of the 'new financial architecture,'" *Cambridge Journal of Economics* 33, no. 4 (2009): 563–80.

5. Michael Pryke, "Looking back on the space of a boom: (Re)developing spatial matrices in the City of London," *Environment and Planning A* 26, no. 2 (1994): 235–64; Rachel Weber, "Selling city futures: The financialization of urban redevelopment policy," *Economic Geography* 86, no. 3 (2010): 251–74; Greta Krippner, *Capitalizing on Crisis* (Cambridge, MA: Harvard University Press, 2011).

6. See Donald Trump and Tony Schwartz, *Trump: The Art of the Deal* (New York: Ballantine Books, 2004).

7. See, for example, David Harvey, *The Urbanization of Capital* (Baltimore: Johns Hopkins University Press, 1985). With some exceptions, Marxist scholars of financialization tend to focus on the "longue durée" of capitalist crisis and transformation, ignoring both short-term development booms and busts and their impact on specific places at specific times.

8. Owner-occupied buildings tend to be smaller and less sensitive to these financial dynamics. See Michael Ball, "Is there an office replacement cycle?," *Journal of Property Research* 20, no. 2 (2003): 173–89.

9. Joe Feagin and Robert Parker, *Building American Cities: The Urban Real Estate Game* (Washington, DC: Beard Books, 2002).

10. Jeffrey Pfeffer and Gerald R. Salancik, *The External Control of Organizations: A Resource Dependence Perspective* (Palo Alto: Stanford University Press, 2003).

11. When developers have been unable to fund new investments or repay debt because of tighter lending standards, some have gone public, effectively securitizing their own portfolios. At the start of the Millennial Boom, only 10 percent of the estimated $3 to $4 trillion in commercial real estate in the United States was owned by publicly traded companies. Stan Ross and Dale Anne Reiss, "Making the connection: Real estate services, mergers, and customers," in Urban Land Institute, *The Changing Face of Commercial Real Estate* (Washington, DC: Urban Land Institute, 1999), 30–35. See also Alan Rabinowitz, *The Real Estate Gamble* (New York: American Management Association, 1980).

12. The full-service model began to appear when the industry reconstituted in the early 1990s following a tumultuous period of bankruptcies and corporate insolvencies. Eamonn D'Arcy and Geoffrey Keogh, "Market maturity and property market behaviour: A European comparison of mature and emergent markets," *Journal of Property Research* 11, no. 3 (1994): 215–35; Anne Haila, "Real estate in global cities: Singapore and Hong Kong as property states," *Urban Studies* 37 (2000): 2241–56; and Michael Ball, *Markets and Institutions in Real Estate and Construction* (London: John Wiley & Sons, 2008) provide excellent descriptions of the maturation of this profession.

13. Charles Bagli, "In city real estate, old clans are shrewd again," *New York Times*, February 9, 2010, A1.

14. Only 13 percent of the large U.S.-based real estate firms maintain global operations. Ashok Bardhan and Cynthia Kroll, "Globalization and the real estate industry: Issues, implications, opportunities" (paper prepared for the Sloan Industry Studies Annual Conference,

Cambridge, MA, April 2007), http://web.mit.edu/sis07/www/kroll.pdf. See also Andrew Wood, "The scalar transformation of the US commercial property development industry: A cautionary note on the limits of globalization," *Economic Geography* 80, no. 2 (2004): 119–40.

15. Bruce Carruthers and Arthur Stinchcombe, "The social structure of liquidity: Flexibility, markets, and states," *Theory and Society* 28 (1999): 353–82, 353.

16. Kevin Fox Gotham, "The secondary circuit of capital reconsidered: Globalization and the U.S. real estate sector," *American Journal of Sociology* 112 (2006): 231–75, 235.

17. Michael Ball, "The 1980s property boom," *Environment and Planning A* 26, no. 5 (1994): 671–95.

18. David Harvey, "The urban process under capitalism: A framework for analysis," *International Journal of Urban and Regional Research* 2, 1–4 (1978): 101–31.

19. Gordon Clark and Kevin O'Connor, "The informational content of financial products and the spatial structure of the global finance industry," in *Spaces of Globalization: Reasserting the Power of the Local*, ed. Kevin Cox (New York: Guilford Press, 1997), 89–114.

20. My use of "life cycle" is not meant to imply that the movement of finance through the built environment is natural or predetermined. Real estate actors in the United States have altered professional practices and lobbied hard for reforms to allow for this particular progression. In other countries, and in other asset classes, finance interfaces with real estate in different ways.

21. An exception are older "real estate families" who often eschew debt and use their own capital to finance deals. Family-owned firms also tend to have longer time horizons: for example, New York–based Douglas Durst assembled land for over four decades before building the fifty-two-story 1 Bryant Park Tower, the largest office tower built in New York City during the Millennial Boom. Bagli, "In city real estate, old clans are shrewd again." See also Tom Shachtman, *Skyscraper Dreams: The Great Real Estate Dynasties of New York* (New York: Little, Brown, 1991).

22. Construction loan interest rates can carry a premium of several percentage points above conforming loan rates.

23. Money center banks hold the majority of the world's assets, maintain a global presence, and engage in wholesale banking with retail banks and large corporations.

24. It is difficult to estimate how much public assistance developers have received for their projects, given the proliferation and variation of sources of such assistance across the United States. One study found that public incentives cost state and local governments an estimated $70 billion annually. See Kenneth Thomas, *Investment Incentives and the Global Competition for Capital* (New York: Palgrave Macmillan, 2010).

25. Or construction lenders convert these loans to "mini-perm" financing if the project is unable to attract a permanent lender or new owner. Short-term, higher-interest "bridge" or "mezzanine" loans are other ways of funding a project that does not yet qualify for longer-term financing.

26. Yuliya Demyanyk and Kent Cherny, "Bank exposure to commercial real estate," *Economic Trends* (Federal Reserve Bank of Cleveland, August 2009).

27. "Large portion of bank loans backed by commercial real estate are not CRE loans," MBA Research Policy Notes (Washington, DC: Mortgage Bankers Association, May 2007), http://www.mbaa.org/files/Research/IssueN02LargePortionofBankLoansBackedbyCommercialRealEstateAreNotCRELoans.pdf.

28. See, for example, the Depository Institutions Deregulation and Monetary Control Act

of 1980 and the Garn–St. Germain Depository Institutions Act of 1982. For accounts of banking deregulation, see Anthony Downs, *The Revolution in Real Estate Finance* (Washington, DC: Brookings Institution, 1985); Daniel Immergluck, *Foreclosed: High-Risk Lending, Deregulation, and the Undermining of America's Mortgage Market* (Ithaca, NY: Cornell University Press, 2009); Brian P. Holly, "Regulation, competition, and technology: The restructuring of the US commercial banking system," *Environment and Planning A* 19 (1987): 633–52; J. Dennis Lord, "Geographic deregulation of the US banking industry and spatial transfers of corporate control," *Urban Geography* 13, no. 1 (1992): 25–48.

29. For example, in 1984 the federal government abolished the withholding tax on foreign holders of bonds issued by U.S. citizens, and in 1994 the Reigle-Neal Interstate Banking and Branching Act allowed banks to operate across state lines.

30. Immergluck, *Foreclosed*; Karl Beitel, "Financial cycles and building booms: A supply side account," *Environment and Planning A* 32 (2000): 2113–32.

31. Philip Ashton, "An appetite for yield: The anatomy of the subprime mortgage crisis," *Environment and Planning A* 41 (2009): 1420–41.

32. The Employee Retirement Security Act of 1974 created incentives for pension funds to invest in CRE through its "full diversification" mandate. See Zvi Bodie, "Pension funds and financial innovation (working paper 3101, National Bureau of Economic Research, 1989), http://www.nber.org/papers/w3101.

33. Efforts to measure average holding periods for real estate have been complicated by selection bias (samples include only properties that have sold) and incomplete information (studies cannot control for the duration of underlying leases or market strength at any one time across locations). David Collett, Colin Lizieri, and Charles Ward, "Timing and the holding periods of institutional real estate," *Real Estate Economics* 31, no. 2 (2003): 205–22; Jeffrey Fisher and Michael Young, "Holding periods for institutional real estate in the NCREIF database," *Real Estate Finance* 17, no. 3 (2000): 27–34.

34. Lingling Wei and Mike Spector, "Tishman venture gives up Stuyvesant project," *Wall Street Journal*, January 25, 2010.

35. Charles Bagli, "Real estate developers prosper despite defaults," *New York Times*, January 1, 2011.

36. These vehicles were established by Congress in 1960 but did not play a large role until the passage of the Tax Reform Act of 1986, the Technical and Miscellaneous Revenue Act of 1988, and the REIT Modernization Act of 1999. These three acts removed regulatory barriers to qualifying for and sustaining REIT status and empowered REITs to own, operate, and manage many different types of income-producing properties.

37. Barney Warf, "Vicious circle: Financial markets and commercial real estate in the United States," in *Money, Power and Space*, ed. Ron Martin (New York: Wiley, 1994), 309–26.

38. Lynn Fisher and Brian Ciochetti, "Determinants of the terms and performance of office leases" (working paper, Real Estate Research Institute, 2007), http://www.reri.org/research/article_pdf/wp133.pdf.

39. Peter Wissoker, "From insurance to investments: Financialisation and the supply side of life insurance and annuities in the USA (1970–2006)," *Cambridge Journal of Regions, Economy and Society* 6, no. 3 (2013): 401–18.

40. Devin Leonard, "How Lehman Brothers got its real estate fix," *New York Times*, May 3, 2009.

41. SL Green, Vornado, and Brookfield. Bagli, "In city real estate, old clans are shrewd again," A1.

42. Mortgages have not been the only asset to undergo this kind of transformation; car loans, student loans, credit card debt, equipment leasing, and manufactured housing have all been converted into saleable instruments.

43. For an exception, see Lynne Sagalyn, "Real estate risk and the business cycle: Evidence from security markets," *Journal of Real Estate Research* 5, no. 2 (1990): 203–19.

44. Goetzmann and Newmann, "Securitization in the 1920's." They sold bonds retail to small investors to raise funds for building, whereas previously only wealthy individuals, banks, and insurance companies invested in real estate.

45. Richard Stanton and Nancy Wallace, "CMBS and the role of subordination levels in the crisis of 2007–2009" (working paper 16206, National Bureau of Economic Research, 2010), http://www.nber.org/papers/w16206.

46. Viral Acharya and Matthew Richardson, "Causes of the financial crisis," *Critical Review* 21, no. 2–3 (2009): 195–210. Some of these instruments take the original mortgages off originators' balance sheets (such as real estate mortgage investment conduits, or REMICs), while others leave them there (such as commercial mortgage obligations, or CMOs).

47. Lingling Wei, "CMBS revival marks step toward recovery," *Wall Street Journal,* September 22, 2010.

48. In similar fashion, public sector debt instruments such as tax increment financing allow developers to receive funds up front from the proceeds of the sale of future property tax streams to investors. This form of "municipal securitization" provides cities with more leveraged funds to lend or invest as equity in real estate. See Weber, "Selling city futures."

49. Conflicts of interest (and a decline in bond quality) are exacerbated when issuers go "ratings shopping," i.e., selecting the ratings agency that offers the most favorable credit assessments.

50. Gotham, "Secondary circuit of capital reconsidered," 262.

51. Anthropologists and others sensitive to the discursive construction of financial narratives remind us that distinctions such as fictional/fundamental, embedded/disembedded, and speculative/productive are dangerous when discussing dimensions of financial value. When an economy is so thoroughly financialized, the links between "underlying" assets and the financial instruments derived from them are tenuous at best. As Janet Roitman notes, "It is meaningless to try to trace wealth back to home values, as some kind of "fundamental" value that could be located or determined outside of this system of financial valuation and production." Roitman, *Anti-crisis* (Durham, NC: Duke University Press, 2013), 54. See also Christian Marazzi, *Capital and Language: From the New Economy to the War Economy* (Los Angeles: Semiotext, 2008); Chris Muellerleile, "Turning financial markets inside out: Polanyi, performativity and disembeddedness," *Environment and Planning A* 45 (2013): 1625–42.

52. Developer A.

53. Chris Macke, "Commentary: Forget 'location, location, location;' It's now 'timing, timing, timing,'" *National Real Estate Investor,* October 20, 2010, http://nreionline.com/finance-amp-investment/commentary-forget-location-location-location-it-s-now-timing-timing-timing.

54. Warf, "Vicious circle," 309.

55. Quoted in Susan Fainstein, *The City Builders: Property, Politics, and Planning in London and New York, 1980–2000,* 2nd ed. (Lawrence: University Press of Kansas, 2001), 67.

56. Harvey, "Urban process under capitalism"; David Harvey, *The Limits to Capital* (Chicago: University of Chicago Press, 1982); David Harvey, *The Urbanization of Capital: Studies in the History and Theory of Capitalist Urbanization* (Baltimore: Johns Hopkins University Press, 1985). See also Beitel, "Financial cycles and building booms"; Gotham, "Secondary circuit of

capital reconsidered"; Igal Charney, "Three dimensions of capital switching within the real estate sector: A Canadian case study," *International Journal of Urban and Regional Research* 25 (2001): 740–58; Jerry Coakley, "The integration of property and financial markets," *Environment and Planning A* 26 (1994): 697–713; Helga Leitner, "Capital markets, the development industry, and urban office market dynamics: Rethinking building cycles," *Environment and Planning A* 26, no. 5 (1994): 779–802; and Christophers, "Revisiting the urbanization of capital," for a more extensive discussion of capital switching.

57. Harvey originally viewed real estate as a last-ditch hope for "finding productive uses for rapidly over-accumulating capital," but he has since acknowledged this sector "as an investment channel in its own right" with distinct logics and autonomy from the rest of the economy. Real estate investment occurs even when profitable opportunities in the primary circuit exist because of property's attractiveness as a hedge against inflation, a means of diversifying corporate portfolios, and a way to shelter taxable income. Thus, switching may occur when yields are *less* than in the primary circuit but risk levels are more acceptable. David Harvey, *Limits to Capital*. See also Joe Feagin, "The secondary circuit of capital: Office construction in Houston, Texas," *International Journal of Urban and Regional Research* 11, no. 2 (1987): 172–92; Anne Haila, "Four types of investment in land and property," *International Journal of Urban and Regional Research* 15, no. 3 (1991): 343–65.

58. Nonrecourse debt is secured by a pledge of collateral, typically real property, but the borrower is not personally liable for it. In the event of default, the lender's recovery is limited to the collateral. Lenders A and E. See also Robert Litan, "Banks and real estate: Regulating the unholy alliance," in *Real Estate and the Credit Crunch*, ed. Lynn Browne and Eric Rosengren (conference series no. 36, Federal Reserve Bank of Boston, 1992), 187–217.

59. Steven Grenadier, "The strategic exercise of options: Development cascades and overbuilding in real estate markets," *Journal of Finance* 51 (1996): 1653–79.

60. Raymond Owens, "Commercial real estate overbuilding in the 1980s: Beyond the hog cycle" (working paper 94-06, Federal Reserve Bank of Richmond, 1994), http://www.richmondfed .org/publications/research/working_papers/1994/pdf/wp94-6.pdf. Tenants are likely to move to new buildings during booms because they too can access more credit and can relocate to more expensive quarters.

61. Litan, "Banks and real estate"; Joe Peek and Eric S. Rosengren, "The role of real estate in the New England credit crunch" (working paper 92-4, Federal Reserve Bank of Boston, 1992).

62. John Logan and Harvey Molotch, *Urban Fortunes: The Political Economy of Place* (Berkeley: University of California Press, 1987), 25.

63. E. P. Davis and H. Zhu, "Bank lending and commercial property cycles: Some cross-country evidence" (working paper no. 150, Bank for International Settlements, 2004).

64. In *Towers of Capital: Office Markets and International Financial Services* (Oxford: Wiley-Blackwell, 2009), Colin Lizieri notes that infrequent trading creates a scarcity of accurate information about property values, making this sector susceptible to both "rational" and speculative pricing bubbles. During booms, however, the increased pace of sales should make more such information available. See Daniel P. McMillen and Rachel Weber, "Thin markets and property tax inequities: A multinomial logit approach," *National Tax Journal* 61 (2008): 653–71.

65. Quoted in Alby Gallun, "Bubble-era cash-out refinancings pay off for real estate investors," *Crain's Chicago Business*, September 20, 2010.

66. Ball, "1980s property boom."

67. Hyman Minsky, *Can "It" Happen Again? Essays on Instability and Finance* (Armonk, NY: M. E. Sharpe, 1984).

68. Donald MacKenzie, "The credit crisis as a problem in the sociology of knowledge," *American Journal of Sociology* 116, no. 6 (May 2011): 1778–1841.

69. Beitel, "Financial cycles and building booms," 2127.

70. Christophers, "Revisiting the urbanization of capital," and Robert Beauregard, "Capital switching and the built environment," *Environment and Planning A* 26 (1994): 715–32, point out the difficulties in empirically validating the construct of capital switching.

71. Mainstream economists measure bubbles in terms of serial price correlation between different markets, which they attribute to speculation rather than a response to supply-demand conditions. See Matthew Higgins and Carol Osler, "Asset market hangovers and economic growth: The OECD during 1984–93," *Oxford Review of Economic Policy* 13, no. 3 (1997): 110–34; Stephen Malpezzi and Susan Wachter, "The role of speculation in real estate cycles," *Journal of Real Estate Literature* 13, no. 2 (2005): 141–64; and John Kling and Thomas E. McCue, "Office building investment and the macroeconomy: Empirical evidence, 1973–1985," *Real Estate Economics* 15, no. 3 (1987): 234–55. Critical scholars generally need less statistical evidence of this correlation to draw similar conclusions: see Robert Brenner, *The Boom and the Bubble: The US in the World Economy* (London: Verso, 2002); Ross King, "Capital switching and the role of ground rent: Switching between circuits and switching between submarkets," *Environment and Planning A* 21, no. 6 (1989): 711–38; Manuel Aalbers, "The financialization of home and the mortgage market crisis," *Competition & Change* 12 (2008): 148–66.

72. Goetzmann and Newmann, "Securitization in the 1920's." After the new bonds entered into circulation en masse, 128 buildings over 230 feet were constructed in New York City between 1929 and 1931.

73. Rabinowitz, *Real Estate Gamble*, 236, claims that 1957–1962 is the first wave of publically traded syndications; 1960–1965 is the time when land development securities took off; 1965–1970 is the period when large corporations developed real estate subsidiaries; and 1968–1974 is dominated by syndicates and REITs.

74. See Richard Barras, "Property and the economic cycle: Building cycles revisited," *Journal of Property Research* 11, no. 3 (1994): 183–97; Beitel, "Financial cycles and building booms"; Leitner, "Capital markets"; Richard Herring and Susan Wachter, "Bubbles in real estate markets" (Zell/Lurie Real Estate Center working paper 402, Wharton School, University of Pennsylvania, 2002), http://realestate.wharton.upenn.edu/research/papers/full/402.pdf. This idea contradicts the notion that the increasing securitization of real estate will bring more discipline to the property sector. Mark Gallagher and Anthony Wood, for example, state that such developments will "lessen the risk of overbuilding and ultimately lead to both shorter and shallower real estate cycles." See "Fear of overbuilding in the office sector: How real is the risk and can we predict it?," *Journal of Real Estate Research* 17, no. 1 (1999): 3–32, 6.

75. Leo Grebler and Leland Burns, "Construction cycles in the United States since World War II," *Real Estate Economics* 10 (1982): 123–51; William Wheaton, "The cyclic behavior of the national office market," *Real Estate Economics* 15 (1987): 281–99; Patrick Wilson and John Okunev, "Spectral analysis of real estate and financial assets markets," *Journal of Property Investment and Finance* 17, no. 1 (1999): 61–74.

76. John Maynard Keynes, *The General Theory of Employment, Interest and Money* (London: Macmillan, 1936); see also Hyman Minsky, *Stabilizing an Unstable Economy* (New Haven, CT: Yale University Press, 2008); Coakley, "Integration of property and financial markets"; Brenner, *Boom and the Bubble.*

77. Karl Case, "The real estate cycle and the economy: Consequences of the Massachusetts boom of 1984–1987," *Urban Studies* 29, no. 2 (1992): 171–83; Ben Bernanke and Mark Gertler,

"Agency costs, net worth, and business fluctuations," *American Economic Review* 79, no. 1 (1989): 14-31; Edward Leamer, "Housing IS the business cycle" (working paper 13428, National Bureau of Economic Research, 2007), http://www.nber.org/papers/w13428.

78. Robert Beauregard, *When America Became Suburban* (Minneapolis: University of Minnesota Press, 2006); Robert Bruegmann, *Sprawl: A Compact History* (Chicago: University of Chicago Press, 2006). Later in this same decade, however, developers were led on by investors' search for tax-sheltered income and by Wall Street firms entering the mortgage brokerage and banking fields. See Rabinowitz, *Real Estate Gamble.*

79. James Freund, Timothy Curry, Peter Hirsch, and Theodore Kelley, "Commercial real estate and the banking crises of the 1980s and early 1990s," in *History of the 80s*, vol. 1, *An Examination of the Banking Crises of the 1980s and Early 1990s* (Washington, DC: Federal Deposit Insurance Corporation, 1997), 137-65; Rabinowitz, *Real Estate Gamble.*

80. Bertrand Renaud, "How real estate contributed to the Thailand financial crisis," in *Asia's Financial Crisis and the Role of Real Estate*, ed. M. Koichi and Bertrand Renaud (London: M. E. Sharpe, 2000), 183-207.

81. Peter Englund, "The Swedish banking crisis: Roots and consequence," *Oxford Review of Economic Policy* 15, no. 3 (1999): 80-97.

82. Richard Herring and Susan Wachter, "Real estate booms and banking busts: An international perspective" (working paper 99-27, Wharton Financial Institutions Center, 1999), http://fic.wharton.upenn.edu/fic/papers/99/9927.pdf. For general models of financial crises, see Franklin Allen and Douglas Gale, "Bubbles and crises," *Economic Journal* 110 (2000): 236-55.

83. Özgür Orhangazi, "Financialisation and capital accumulation in the non-financial corporate sector: A theoretical and empirical investigation on the US economy: 1973-2003," *Cambridge Journal of Economics* 32 (2008): 863-86.

84. Paul Langley, "The making of investor subjects in Anglo-American pensions," *Environment and Planning D* 24 (2006): 919-34; Gordon Clark, *Pension Fund Capitalism* (Oxford: Oxford University Press, 2000).

85. Adam Tickell, "Dangerous derivatives: Controlling and creating risks in international money," *Geoforum* 31 (2000): 87-99; Michael Pryke and John Allen, "Monetized time-space: Derivatives—Money's 'new imaginary'?," *Economy and Society* 29 (2000): 264-84.

86. Deniz Igan and Marcelo Pinheiro, "Exposure to real estate in bank portfolios," *Journal of Real Estate Research* 32 (2010): 47-74.

87. Bertrand Renaud, "The 1985 to 1994 global real estate cycle: An overview," *Journal of Real Estate Literature* 5, no. 1 (1997): 13-44.

88. The ratio of real estate foreign direct investment to total real estate GDP increased from 0.44 in 1973 to 4.21 in 1987. Gotham, "Secondary circuit of capital reconsidered."

89. Garner, "Is commercial real estate reliving the 1980s and early 1990s?"

90. Alireza Dehesh and Cedric Pugh, "Property cycles in a global economy," *Urban Studies* 37, no. 13 (2000): 2581-2602; Simon Guy and Robert Harris, "Property in a risk society: Towards marketing research," *Urban Studies* 34, no. 1 (1997): 125-40.

91. Warf, "Vicious circle."

92. Thanks to Peter Wissoker for pointing this out.

93. Lynn Browne and Karl E. Case, "How the commercial real estate boom undid the banks," *Federal Reserve Bank of Boston Conference Series: Proceedings* 36 (1992): 57-113.

94. Marc Weiss, "The politics of real estate cycles," *Business and Economic History* 20, no. 2 (1991): 127-35.

95. Browne and Case, "How the commercial real estate boom undid the banks."

96. Garner, "Is commercial real estate reliving the 1980s and early 1990s?"

97. David Birch, Susan M. Jain, William Parsons, and Zhu Xiao Di, *America's Future Office Space Needs* (Washington, DC: National Association of Industrial and Office Parks, 1990).

98. Joel Warren Barna, *The See-Through Years: Creation and Destruction in Texas Architecture and Real Estate, 1981–1991* (Houston: Rice University Press, 1992).

99. Browne and Case, "How the commercial real estate boom undid the banks."

100. Warf, "Vicious circle."

101. Patric Hendershott and Edward J. Kane, "Causes and consequences of the 1980s commercial construction boom," *Journal of Applied Corporate Finance* 5, no. 1 (1992): 61–70.

102. Glenn R. Mueller, "Predicting long-term trends and market cycles in commercial real estate," *Wharton Real Estate Review* 5 (2001): 14.

103. Bardhan and Kroll, "Globalization and the real estate industry."

104. Krippner, *Capitalizing on Crisis.*

105. Garner, "Is commercial real estate reliving the 1980s and early 1990s?"

106. Adam Levitin and Susan Wachter, "The commercial real estate bubble," *Harvard Business Law Review* 3 (2013): 83–118.

107. REITs and insurance companies had more restrictions on leverage. While the public market capitalization of all REITs was $438 billion in 2007, securitized multifamily and commercial mortgages constituted almost twice that amount ($907 billion). Commercial Mortgage Securities Association, "The role of Commercial Mortgage Backed Securities (CMBS) and Commercial Real Estate Collateralized Debt Obligations (CRE CDOs) in the U.S. real estate finance market," accessed January 18, 2015, https://www.crefc.org/uploadedFiles/CMSA_Site_Home/Industry_Resources/Research/Industry_Overview/CMBS%20Summary(1).pdf.

108. See Aalbers, "Financialization of home"; Ashton, "Appetite for yield"; Immergluck, *Foreclosed.*

109. Demyanyk and Cherny, "Bank exposure to commercial real estate."

110. Steve Bergsman, "Sweet synergies: JPMorgan's CMBS business shot up the issuer rankings following merger with Bank One," *Mortgage Banking* 66, no. 4 (2006): 46–49.

111. Garner, "Is commercial real estate reliving the 1980s and early 1990s?"

112. Robert Cropf and G. Wendel, "The determinants of municipal debt policy: A pooled time-series analysis," *Environment and Planning C* 16 (1998): 211–24; Weber, "Selling city futures."

113. U.S. Bureau of the Census, "Census of Governments: Finance—Surveys of State and Local Government Finances," 1997 and 2008.

114. Author's calculation based on "Current Employment Statistics," 2000–2010, U.S. Bureau of Labor Statistics, http://www.bls.gov/ces/. FIRE employment in the United States peaked in 2003 at 6.14 percent of total employment and decreased to 5.93 percent in 2009. These shares, however, vary considerably across major office markets, with 9.14 percent of workers employed in the FIRE sector in New York City compared with 4.56 percent of workers in Washington, DC (in 2008).

115. Stanely Aronowitz, *Just around the Corner: The Paradox of the Jobless Recovery* (Philadelphia: Temple University Press, 2005); Mary Daly, Bart Hobijn, and Joyce Kwok, "Jobless recovery redux?," *Economic Letter* 18 (Federal Reserve Bank of San Francisco, 2009), http://www.frbsf.org/economic-research/publications/economic-letter/2009/june/jobless-recovery/.

116. David Harvey, *The Enigma of Capital* (New York: Oxford University Press, 2010), 17.

117. Ashok Bardhan and Richard Walker, "California shrugged: Fountainhead of the Great Recession," *Cambridge Journal of Regions, Economy and Society* 4, no. 3 (2011): 303–22.

118. U.S. Bureau of the Census, "Value of Construction Put in Place," 2009, http://www.census.gov/const/www/c30index.html.

119. Garner, "Is commercial real estate reliving the 1980s and early 1990s?"

120. National Council of Real Estate Investment Fiduciaries, "The NCREIF Property Index: Trends, transaction and appraisal capitalization rates," 1992–2008. See also "The recovery . . . was that it?," NCREIF Research Committee Webinar, August 23, 2011, https://www.ncreif.org/webinar.aspx.

121. Lingling Wei and Peter Grant, "Commercial real estate lurks as next potential mortgage crisis," *Wall Street Journal*, August 31, 2009.

122. See Ronen Elul, "Securitization and mortgage default" (working paper 09-21/R, Federal Reserve Bank of Philadelphia, 2011), http://www.philadelphiafed.org/research-and-data/publications/working-papers/2009/wp09-21R.pdf; and John Krainer and Elizabeth Laderman, "Mortgage loan securitization and relative loan performance," *Journal of Financial Services Research* 45, no. 1 (2014): 39–66, for comparable findings in the case of residential mortgage securitization.

123. Sam Chandan, "The past, present, and future of CMBS," *Wharton Real Estate Review*, Spring 2012, http://realestate.wharton.upenn.edu/review/index.php?article=236.

124. See chapter 1, table 1.1; Mark Heschmeyer, "U.S. CMBS REOs top $11 billion; one-third of all delinquencies;" *CoStar News*, May 16, 2012, http://www.costar.com/News/Article/US-CMBS-REOs-Top-$11-Billion;-One-Third-of-All-Delinquencies/138499.

125. Chandan, "Past, present and future of CMBS."

126. Quoted in Charles Bagli, "Failed deals replace boom in New York real estate," *New York Times*, October 1, 2008, B1.

127. CRE Finance Council, "Sales of large commercial properties," in *Compendium of Statistics* (New York: CREFC, October 11, 2011), http://www.crefc.org/uploadedFiles/CMSA_Site_Home/Industry_Resources/Research/Industry_Statistics/CMSA_Compendium.pdf.

128. Kathryn Byun, "The U.S. housing bubble and bust: Impacts on employment" *Monthly Labor Review*, December 2010, 10, http://www.bls.gov/opub/mlr/2010/12/art1full.pdf.

129. Ibid.

130. Developer C.

131. Weber, "Selling city futures." See also Josh Pacewicz, "Tax increment financing, economic development professionals and the financialization of urban politics," *Socio-Economic Review* 11 (2012): 413–40.

Chapter Three

1. William Mazzarella, *Shoveling Smoke: Advertising and Globalization in Contemporary India* (Durham, NC: Duke University Press, 2005), 47.

2. Michel Callon, Cécile Méadel, and Vololona Rabeharisoa, "The economy of qualities," *Economy and Society* 31, no. 2 (2002): 194–217, 203.

3. "Replacement demand" is the term used to describe demand that results from occupants' existing space becoming unusable because of physical deterioration or "obsolescence." Nathan Schloss, "Use of employment data to estimate office space demand," *Monthly Labor Review* 107 (1984): 40. See also Michael Ball, "Is there an office replacement cycle?," *Journal of Property Research* 20, no. 2 (2003): 173–89.

4. In Richard Tucker, ed., *The Marx-Engels Reader* (New York: Norton, 1978), 303.

5. Annelise Riles, *Collateral Knowledge: Legal Reasoning in the Global Financial Markets* (Chicago: University of Chicago Press, 2011).

6. In lease negotiations, tenants encounter the building managers, most often working out of the largest brokerage houses and development firms, rather than the owners themselves.

7. "Underneath variably liquid financial claims lies the use value of something, or a going concern that produces real goods and services." Bruce Carruthers and Arthur Stinchcombe, "The social structure of liquidity: Flexibility, markets, and states," *Theory and Society* 28, no. 3 (1999): 353–82, 359.

8. Jeffrey Hornstein, *A Nation of Realtors: A Cultural History of the Twentieth-Century American Middle Class* (Durham, NC: Duke University Press, 2005).

9. Ibid.

10. In an era when many of these data are online, brokers must justify their commissions based on the other skills (e.g., customer service) they bring to the table. In other words, the "soft" expertise associated with sociability and affect (earlier works referred to such "unteachable" qualities as "character" and "tact"; see Hornstein, *Nation of Realtors*) as well as local knowledge still defines a broker's role.

11. Broker G.

12. The social linkages within and between professional communities encourage emulation, groupthink, and what Paul DiMaggio and Walter Powell call "normative isomorphism." "The iron cage revisited: Institutional isomorphism and collective rationality in organizational fields," *American Sociological Review* 48 (1983): 147–60; see also Karen Ho, *Liquidated: An Ethnography of Wall Street* (Durham, NC: Duke University Press, 2009); Sarah Hall, "Financialised elites and the changing nature of finance capitalism: Investment bankers in London's financial district," *Competition & Change* 13, no. 2 (2009): 173–89; Donald MacKenzie, "Social connectivities in global financial markets," *Environment and Planning D* 22, no. 1 (2004): 83–102; Bruce Carruthers and Jeong-Chul Kim, "The sociology of finance," *Annual Review of Sociology* 37 (2011): 239–59; Bruce Carruthers, "A sociology of bubbles," *Contexts* 8, no. 3 (2009): 22–26.

13. Broker C and Facilities Manager A.

14. Amanda Robert, "Lawyers and real estate experts talk office space," *Chicago Lawyer*, March 4, 2011.

15. Mazzarella, *Shoveling Smoke*, 167.

16. Broker F.

17. Callon, Méadel, and Rabeharisoa, "Economy of qualities."

18. Consultant A.

19. Broker A.

20. A 1924 article in *Architectural Forum* mentions this classificatory system for office buildings. Clarence Coley, "Office buildings, past, present and future," September, 113–14. Thanks to Jean Guarino for pointing this out.

21. "Building Class Definitions," Building Owners and Managers Association International, accessed December 23, 2014, http://www.boma.org/research/Pages/building-class-definitions.aspx.

22. Investment Broker A.

23. Broker F. See also Leslie Sklair, "The transnational capitalist class and contemporary architecture in globalizing cities," *International Journal of Urban and Regional Research* 29, no. 3 (2005): 485–500.

24. In the absence of a governing body or formal designation process, each major brokerage firm independently determines a building's class. "There is a universal desire for a classification system that is somewhat objective, but unfortunately, it is really a very subjective process," noted the executive vice president of a global brokerage firm. Quoted in Julie Satow, "Class-consciousness in the office building market," *New York Times*, July 26, 2011.

25. Carruthers and Stinchcombe, "Social structure of liquidity."

26. Broker F.

27. John Stuart Mill, *Principles of Political Economy* (London: John W. Parker, 1848).

28. Caitlin Zaloom, *Out of the Pits: Traders and Technologies from Chicago to London* (Chicago: University of Chicago Press, 2006); Mary Poovey, *A History of the Modern Fact: Problems of Knowledge in the Sciences of Wealth and Society* (Chicago: University of Chicago Press, 1998); Theodore Porter, *Trust in Numbers: The Pursuit of Objectivity in Science and Public Life* (Princeton, NJ: Princeton University Press, 1996).

29. One of the first handbooks published in the United States, *McMichael's Appraising Manual*, begins with this revealing paragraph: "There were many men, in the past, who had only one basis for the valuation estimates they made, and that was: 'my opinion.' It was their last line of defense, and stood as an impregnable wall against assaults from all quarters. There were other men who felt that an opinion must be based on something in large part explainable; that a scientific approach should be taken in appraisal of real estate; and that there were fundamental concepts and measurable qualities involved. The latter group lived through periods of serious criticism and were the true pioneers." Stanley McMichael (Englewood Cliffs, NJ: Prentice Hall, 1931).

30. They also started the American Institute of Real Estate Appraisers in 1932 to provide the education for state certification and promulgate rules and standards for the profession.

31. The appraisal industry relied heavily on racial and ethnic ranking systems to establish values. In 1977 the Justice Department sued appraisers and lenders to discontinue these practices, but until then appraisers unquestioningly considered "undesirable racial elements" in an area to be a leading cause of depreciation. Jennifer Light, "Nationality and neighborhood risk at the origins of FHA underwriting," *Journal of Urban History* 36, no. 5 (2010): 634–71.

32. Appraiser E.

33. Appraiser A.

34. Appraiser B.

35. Banks initially had "appraisal committees" composed of their lending staff, then formed separate departments of appraisers. Because of the temptation for loan officers to skew values to suit the bank's objectives, lenders now rely more on the opinions of independent contract ("fee") appraisers rather than on those of their own staff.

36. Appraiser A.

37. The word first appears in the *Oxford English Dictionary* in 1828. It was first used frequently in the field of biology to mean "gradual disappearance of an organ as a consequence of disuse." Bruce Mazlish, "Obsolescence and 'obsolescibles' in planning for the future," in *Planning for Diversity and Choice*, ed. Stanford Anderson (Cambridge, MA: MIT Press, 1968), 156.

38. Neil Harris, *Building Lives* (New Haven, CT: Yale University Press, 1999).

39. Organizations such as the National Association of Building Owners and Managers (what became the Building Owners and Managers Association, or BOMA) initiated a series of investigations into the problem, commissioning studies of the demolitions of the Tacoma Building in Chicago and the Gillender Building in Manhattan.

40. Eventually, postwar critics of "planned obsolescence" in the production of consumer goods such as Marshall McLuhan, John Kenneth Galbraith, and Victor Papanek gave stronger voice to this sentiment and pointed out what they saw as an unhealthy alliance between producers and advertisers selling mass-produced products designed to be replaced every few years. Giles Slade, *Made to Break: Technology and Obsolescence in America* (Cambridge, MA: Harvard University Press, 2009).

41. Maurice Moonitz, "The risk of obsolescence and the importance of the rate of interest," *Journal of Political Economy* 51, no. 4 (August 1943): 348–55.

42. From the *Century Dictionary*; quoted in Earle Shultz, "The effect of obsolescence on the useful and profitable life of office buildings," *Buildings and Building Management*, August 7, 1922, 13–23.

43. Most textbooks define two kinds of obsolescence: the "economic" obsolescence described above, and "functional" obsolescence, which refers to those qualities endemic to the property that could cause it to lose value. For example, faulty design would doom a product to an early grave. The broader term "depreciation" is sometimes used to describe devaluation that may be caused either by obsolescence or by physical deterioration.

44. Appraisers developed the concept of a parcel's "highest and best use," i.e., the use that would produce its highest value, regardless of the parcel's actual current use, that would also be legally, physically, and financially feasible. The concept is discussed in one of the first appraisal textbooks: "The building which provides the greatest amount of economic rent from the ground is the building which is fitted to the highest and best use of that ground." Frederick Babcock, *The Appraisal of Real Estate* (New York: Macmillan, 1924).

45. J. L. Jacobs, "Neighborhood and property obsolescence in the assessment process," *Journal of Land & Public Utility Economics* 17, no. 3 (1941): 344–53, 353.

46. Babcock, *Appraisal of Real Estate*.

47. Shultz, "Effect of obsolescence."

48. Ibid.

49. Ibid.; Andrew E. Baum, "Quality, depreciation and property performance," *Journal of Real Estate Research* 8, no. 4 (1993): 541–65; David Brazell, Lowell Dworin, and Michael Walsh, *A History of Federal Tax Depreciation Policy* (Washington, DC: Office of Tax Analysis, U.S. Treasury Department, 1989). They emulated the English income tax law of 1897, which pioneered the use of obsolescence as grounds for claiming deductions. See Harris, *Building Lives*.

50. Proponents of the "life cycle" theory of neighborhoods believe that some areas have to be avoided for decades to allow complete abandonment and eventual redevelopment to occur. Edgar Hoover and Raymond Vernon, *Anatomy of a Metropolis* (Garden City, NY: Doubleday Anchor, 1959).

51. Appraiser A.

52. Daniel Abramson, "Obsolescence: Notes toward a history," *Praxis: Journal of Writing Building* 5 (2003): 106–12, 109.

53. Transitions to new commercial districts are often contested by residents and businesses that benefit from older spatial matrices or who are restricted by discrimination or inadequate resources from trading up. Susan Fainstein reminds us that much of what constitutes urban politics arises from such contestations. Fainstein, *The City Builders: Property, Politics, and Planning in London and New York, 1980–2000*, 2nd ed. (Lawrence: University Press of Kansas, 2001).

54. Claus Offe, *Contradictions of the Welfare State* (Cambridge, MA: MIT Press, 1984), 124.

55. Shultz, "Effect of obsolescence."

56. Membership organizations of appraisers and brokers fought for one such intervention, zoning ordinances, in the years following World War I. By clustering like uses and expelling unlike ones, zoning could enhance property values and was worth the cost of its infringement on individual liberties and property rights. Marc A. Weiss, *The Rise of the Community Builders: The American Real Estate Industry and Urban Land Planning* (New York: Columbia University Press, 2002).

57. Robert Fogelson, *Downtown: Its Rise and Fall, 1880–1950* (New Haven, CT: Yale University Press, 2001).

58. In *The City Builders*, Fainstein problematizes the notion of interest formation. She argues that capitalists pursue their economic interests through state action, but she views these interests as neither held in common nor predetermined. Instead, they arise in the context of varied circumstances of local dependency wherein reigning economic, political, and ideological factors figure into the municipality's decisions to favor certain capitalists over others. See also John Mollenkopf, *The Contested City* (Princeton, NJ: Princeton University Press, 1983).

59. Josh Pacewicz, "Tax increment financing, economic development professionals and the financialization of urban politics," *Socio-Economic Review* 11 (2012): 413–40.

60. The overlapping nature of the different professions is reflected by the fact that they participate in the same membership organizations (American Planning Association, Urban Land Institute, the Council of Development Finance Agencies, and the International Economic Development Council), which do not distinguish between public and private employers. Rachel Weber and Sara O'Neil-Kohl, "The historical roots of tax increment financing, or how real estate consultants kept urban renewal alive," *Economic Development Quarterly* 27, no. 3 (2013): 193–207.

61. Harvey Molotch, "The city as a growth machine," *American Journal of Sociology* 82, no. 2 (1976): 309–32; Mollenkopf, *Contested City*; Kevin Cox and Andrew Mair, "Locality and community in the politics of local economic development," *Annals of the Association of American Geographers* 78 (1988): 307–25; Helga Leitner, "Cities in pursuit of economic growth: The local state as entrepreneur," *Political Geography Quarterly* 9, no. 2 (1990): 146–70; Elizabeth Strom, "Rethinking the politics of downtown development," *Journal of Urban Affairs* 30, no. 1 (2013): 37–61.

62. Despite a structural bias toward capital and external pressure to pursue growth, cities undertake redevelopment schemes inconsistently. Some cities take up this mantle vigorously, seeing large-scale projects as the most effective means of competing with the suburbs and stemming decline. Others are more ambivalent or cautious. And in still other cases, local governments seek to pursue aggressive redevelopment agendas but are stymied by political opposition or the high cost of land. Fainstein, *City Builders*.

63. Rachel Weber, "Extracting value from the city: Neoliberalism and urban redevelopment," *Antipode* 34 (2002): 519–40.

64. Christopher Mele, *Selling the Lower East Side: Culture, Real Estate, and Resistance in New York City* (Minneapolis: University of Minnesota Press, 2000); Mark Gottdiener, *The Theming of America: Dreams, Visions, and Commercial Spaces* (Boulder, CO: Westview Press, 1997).

65. David Harvey notes that Fordist strategies for solving the overaccumulation problem included "savage devaluation," referencing the mass torching of surplus coffee plants that occurred in Brazil in the 1930s and 1940s. Harvey, *The Condition of Postmodernity* (Malden, MA: Blackwell, 1990), 181.

66. Neil Smith, *The New Urban Frontier: Gentrification and the Revanchist City* (New York: Routledge, 1996).

67. Tim Edensor, *Industrial Ruins: Space, Aesthetics and Materiality* (Oxford: Berg, 2005).

68. Smith, *New Urban Frontier*, 192.

69. Harvey, however, focused on private activity, particularly the decisions of landlords in Baltimore in the 1970s to remove rental units from circulation in order to boost returns. State action in the service of capital can perform this same function. "Class-monopoly rent, finance capital and the urban revolution," *Regional Studies* 8, no. 3–4 (1974): 239–55.

70. Robert Beauregard, *Voices of Decline: The Post-War Fate of U.S. Cities* (New York: Routledge, 1993).

71. Ibid.; Fogelson, *Downtown*; Max Page, *The Creative Destruction of Manhattan, 1900–1940* (Chicago: University of Chicago Press, 1999); Martin Gold and Lynne Sagalyn, "The use and abuse of blight in eminent domain," *Fordham Urban Law Journal* 38, no. 4 (May 2011): 1119–73.

72. State constitutions require that taxing districts restrict public money to "public purposes." From the urban renewal period on, a finding of blight allowed local planners to satisfy the public purpose test and to acquire and dispose of devalued properties.

73. Clarence Arthur Perry, *The Rebuilding of Blighted Areas: A Study of the Neighborhood Unit in Replanning and Plot Assemblage* (New York: Regional Plan Association, 1933), 9; see also Fogelson, *Downtown*.

74. Benjamin Quinones, "Redevelopment redefined: Revitalizing the central city with resident control," *University of Michigan Journal of Legal Reform* 689 (1993–1994), 689–734.

75. For example, the Chicago Plan Commission's 1942 map of Chicago included "percentage of Negroes" as one of three indicators of blight. Even though buildings on the black South Side were not as old as those on the North and West Sides of the city, they were more frequently categorized as unfit or substandard. D. Bradford Hunt, *Blueprint for Disaster: The Unraveling of Chicago Public Housing* (Chicago: University of Chicago Press, 2009).

76. Beauregard, *Voices of Decline*.

77. Cited in Fogelson, *Downtown*.

78. Gold and Sagalyn, "Use and abuse of blight." Interestingly, the criteria for blight referred to public health statistics, such as death rates from tuberculosis and syphilis, and in some cases were developed by the American Public Health Association. Colin Gordon, "Blighting the way: Urban renewal, economic development, and the elusive definition of blight," *Fordham Urban Law Journal* 31, no. 2 (2003): 305–37.

79. Joel Rast, "Creating a unified business elite: The origins of the Chicago Central Area Committee," *Journal of Urban History* 37, no. 4 (2011): 583–605.

80. Within the United States, the urban renewal period ran from the 1940s through the 1970s and is associated with the Housing Act of 1949 and the amending act of 1954. It is considered to be one of the most robust examples of federal urban policy. John Teaford, "Urban renewal and its aftermath," *Housing Policy Debate* 11, no. 2 (2000): 443–65.

81. Homer Hoyt noted that "blighted areas" were more difficult to redevelop because "obsolete improvements and diversified ownership" made them less competitive than "tracts of virgin prairie" newly accessible by transit and automobile. Hoyt, *One Hundred Years of Land Values in Chicago*, 347.

82. Peter Hall, *Cities of Tomorrow* (New York: Blackwell, 1988).

83. Richard Briffault, "The most popular tool: Tax increment financing and the political economy of local government," *University of Chicago Law Review* 77 (2010): 65–95. See also Weber and O'Neil-Kohl, "Historical roots of tax increment financing."

84. Taxpayers in the district pay real estate taxes on the value of their property prior to the creation of the TIF district as well as on any increase in its value. However, for the life span of the district (about twenty years in most states), all taxes on any new value in the district are directed into a fund to pay for public redevelopment expenditures, such as debt service on bonds floated for infrastructure or private acquisition costs. For more details about the mechanics of TIF, see Rachel Weber, "TIF in theory and practice," in *Financing Economic Development in the 21st Century*, ed. Sammis White and Zenia Kotval (New York: Sage, 2012), 283–301.

85. Robert Beauregard, "The textures of property markets: Downtown housing and office conversions in New York City," *Urban Studies*, 42, no. 13 (2005): 2431–45.

86. Bernard Frieden and Lynne Sagalyn, *Downtown, Inc.: How America Rebuilds Cities* (Cambridge, MA: MIT Press, 1989), 23.

87. See, for example, Andrew Haughwout, "Public infrastructure investments, productivity and welfare in fixed geographic areas," *Journal of Public Economics* 83, no. 3 (2002): 405–28.

88. The historical use of racial criteria in residential lending exposes the power of underwriting standards to give material form to deep-seated prejudices and perpetuate social exclusion. See, for example, Daniel Immergluck, "Redlining redux: Black neighborhoods, black-owned firms, and the regulatory cold shoulder," *Urban Affairs Review* 38 (2002): 22–41; Guy Stuart, *Discriminating Risk: The US Mortgage Lending Industry in the Twentieth Century* (Ithaca, NY: Cornell University Press, 2003); Philip Ashton, "An appetite for yield: The anatomy of the subprime mortgage crisis," *Environment and Planning A* 41 (2009): 1420–41; Manuel Aalbers, "The sociology and geography of mortgage markets: Reflections on the financial crisis," *International Journal of Urban and Regional Research* 33, no. 2 (2009): 281–90.

89. Kerry Vandell and Jonathan Lane, "The economics of architecture and urban design: Some preliminary findings," *Real Estate Economics* 17 (1989): 235–60.

90. Gordon Clark, *Pension Fund Capitalism* (New York: Oxford University Press, 2000).

91. Pension Fund Advisor A.

92. John Logan and Harvey Molotch, *Urban Fortunes: The Political Economy of Place* (London: University of California Press, 1987).

93. Elizabeth Blackmar, "Of REITs and rights: Absentee ownership in periphery," in *City, Country, and Empire: Landscapes in Environmental History*, ed. Jeffry Diefendord and Kirk Dorsey (Pittsburgh, PA: University of Pittsburgh Press, 2005), 81–98.

94. Kevin Fox Gotham, "The secondary circuit of capital reconsidered: Globalization and the U.S. real estate sector," *American Journal of Sociology* 112 (2006): 231–75; Carruthers and Stinchcombe, "Social structure of liquidity."

95. Christopher Leinberger, "The need for alternatives to the nineteen standard real estate product types," *Places* 17 (2005): 24–29.

96. Christopher Leinberger, "Financing progressive development," *Capital Xchange Journal* (May 2001): 7, http://chrisleinberger.com/docs/By_CL/FinancingProgressiveDev.pdf.

97. Broker D.

98. Pension Fund Advisor A and Lender A. It is also due to the fact that the mortgage interest deduction is larger for newer and more expensive buildings.

99. Carol Willis, *Form Follows Finance: Skyscrapers and Skylines in New York and Chicago* (New York: Princeton Architectural Press, 1995).

100. William Goetzmann and Frank Newmann, "Securitization in the 1920's" (working paper 15650, National Bureau of Economic Research, 2010), http://www.nber.org/papers/w15650.pdf.

101. In contrast, some economists believe there is a timeless standard for determining a building's financial efficiency. See Peter Colwell and M. Shahid Ebrahim, "A note on the optimal design of an office building," *Journal of Real Estate Research* 14, no. 2 (1997): 169–74.

102. Baum, "Quality, depreciation and property performance."

103. Architect A.

104. Lender D.

105. Donovan Rypkema, *The Investor Looks at an Historic Building* (Washington, DC: National Trust for Historic Preservation, 1991).

106. Developer D.

107. Such assessments are typically based more in perception than reality, as older buildings can save on utility costs. See Matthew Kahn, Nils Kok, and John M. Quigley, "Carbon emissions from the commercial building sector: The role of climate, quality, and incentives," *Journal of Public Economics* 113 (2014): 1–12.

108. Alan Rowley, "Private property decision makers and the quality of urban design," *Journal of Urban Design* 3, no. 2 (1998): 151–73.

109. Lender A.

110. Bryan Ciochetti and Jeffrey Fisher, "The characteristics of commercial real estate holding period returns" (working paper, Real Estate Research Institute, 2002), http://reri.org/research/article_pdf/wp104.pdf; Geoffrey Rubin, "Is housing age a commodity? Hedonic price estimates of unit age," *Journal of Housing Research* 4, no. 1 (1993): 165–84; G. Stacy Sirmans, John Benjamin, and Emily Zietz, "The environment and performance of industrial real estate," *Journal of Real Estate Literature* 11, no. 3 (2003): 279–324. The conventional wisdom is challenged by other research that finds no clear relationship between building age and utility. See Baum, "Quality, depreciation and property performance."

111. Gordon Clark and Kevin O'Connor, "The informational content of financial products and the spatial structure of the global finance industry," in *Spaces of Globalization: Reasserting the Power of the Local*, ed. Kevin Cox (New York: Guilford Press, 1997), 89–114.

112. Qualities can be reconfigured. Over time and lacking the requisite reinvestment, for example, Class A buildings may be downgraded. See Callon, Méadel, and Rabeharisoa, "Economy of qualities."

113. Mazzarella, *Shoveling Smoke*, 33.

114. Sociologists like Callon would say that cycles are "performative" in the same way that other conceptions of the economy construct possibilities for their own value and market exchange. For example, theories of cycles are not just abstract representations or observations of property markets ambling through time; the theories themselves have independent power to constitute and influence those same markets. The predictive certainty implied by the construct of a cycle bolsters participants' sense of confidence, which encourages risk taking and creates the conditions for the cycle's own momentum and apotheosis. Michel Callon, "What does it mean to say that economics is performative?" (CSI working paper 6, Centre de Sociologie de l'Innovation, 2006); Michel Callon, ed., *The Laws of the Markets* (Oxford: Wiley-Blackwell, 1998), 2; see also Donald MacKenzie and Yuval Millo, "Constructing a market, performing theory: The historical sociology of a financial derivatives exchange," *American Journal of Sociology* 109, no. 1 (2003): 107–45; Donald MacKenzie, Fabian Muniesa, and Lucia Siu, eds., *Do Economists Make Markets? On the Performativity of Economics* (Princeton, NJ: Princeton University Press, 2007).

115. DiMaggio and Powell, "Iron cage revisited."

116. Christian Marazzi, *Capital and Language: From the New Economy to the War Economy* (Los Angeles: Semiotext, 2008). See also Robert Shiller, *Irrational Exuberance* (New York: Random House, 2005), for his notion of a self-fulfilling prophecy driven by the overconfidence of investors.

117. As Keynes has noted, most investors will follow the herd, but a few mavericks will try to outwit and bet against the crowd. Knowledge gaps will affect the spread between buying and selling prices and will create just enough disparity and volatility in values to keep markets buoyant.

118. Broker A.

119. Planner B.

120. Ibid.

Chapter Four

1. Amanda Robert, "Lawyers and real estate experts talk office space," *Chicago Lawyer*, March 4, 2011.

2. Quoted in Thomas Corfman, "Zell considering building new tower in W. Loop," *Crain's Chicago Business*, June 6, 2003.

3. Quoted in Alby Gallun, "Road to riches: John Buck reshapes Wacker Drive," *Crain's Chicago Business*, October 31, 2005.

4. Ibid.

5. Skadden Arps moved from 333 West Wacker (built in 1983 by Buck) to 155 North Wacker (delivered by Buck in 2009).

6. Periodizations of building booms are, by nature, rough. I base mine on new office square footage as well as the number of buildings under construction and delivered in each year. Years were included in a boom period if they witnessed a substantial number of office building deliveries and were preceded and followed by a year that experienced a similar volume of square footage. A single building delivered when the previous and following years were relatively inactive did not constitute a boom.

7. The different real estate data companies divide the smaller commercial and use-specific submarkets along similar, but not always exactly the same, boundaries. CoStar, for example, breaks the Central Area into the River North and North Michigan submarkets, which are separated from the West Loop, Central Loop, and East Loop submarkets by the Chicago River.

8. Robert Lewis, *Chicago Made: Factory Networks in the Industrial Metropolis* (Chicago: University of Chicago Press, 2008).

9. Ibid., 6.

10. Data on new construction in the Loop during these booms were derived primarily from Frank Randall and John Randall's *History of the Development of Building Construction in Chicago* (Champaign: University of Illinois Press, 1999), which provides an encyclopedic wealth of structural and architectural details about Chicago buildings constructed and demolished since the Great Chicago Fire. I also relied on Carl Condit's books on the Chicago School and its favored architectural styles and building technology: *The Chicago School of Architecture: A History of Commercial and Public Building in the Chicago Area, 1875–1925* (Chicago: University of Chicago Press, 1964); *Chicago 1910–1929: Building, Planning, and Urban Technology* (Chicago: University of Chicago Press, 1973); *Chicago 1930–1970: Building, Planning, and Urban Technology* (Chicago: University of Chicago Press, 1974). Thanks to Jean Guarino for compiling this information.

11. Condit, *Chicago School of Architecture.*

12. Ibid.; Carol Willis, *Form Follows Finance: Skyscrapers and Skylines in New York and Chicago* (New York: Princeton Architectural Press, 1995).

13. Homer Hoyt, *One Hundred Years of Land Values in Chicago: The Relationship of the Growth of Chicago to the Rise of Its Land Values, 1830–1933* (Chicago: University of Chicago Press, 1933).

14. Willis, *Form Follows Finance*; see also Alan Rabinowitz, *The Real Estate Gamble* (New York: American Management Association, 1980).

15. Jean Guarino, "Tearing down Chicago's Loop: What caused the demolitions of the 1930s?" (unpublished paper, University of Illinois at Chicago, June 22, 2010).

16. Hoyt, *One Hundred Years of Land Values in Chicago.*

17. Dominic Pacyga, *Chicago: A Biography* (Chicago: University of Chicago Press, 2009).

18. Larry Bennett, *The Third City: Chicago and American Urbanism* (Chicago: University of Chicago Press, 2010), 40.

19. Gerald Manners, "The office in metropolis: An opportunity for shaping metropolitan America," *Economic Geography* 50, no. 2 (1974): 93–110.

20. Floor area ratios (FAR) are used in zoning to determine the allowable density of a building on a particular site. Ratios are generated by dividing the building area by the parcel area. D. Bradford Hunt and Jon DeVries, *Planning Chicago* (Chicago: APA Planners Press, 2013).

21. Ross Miller, *Here's the Deal: The Buying and Selling of a Great American City* (New York: Knopf, 1996).

22. Philip Israelevich and Ramamohan Mahidhara, "Hog butchers no longer: Twenty years of employment change in Metropolitan Chicago," *Economic Perspectives* (Federal Reserve Bank of Chicago) 14, no. 2 (1991): 15–22.

23. Planner A. For comparable examples in Houston and Dallas, see Joel Barna, *The See-Through Years: Creation and Destruction in Texas Architecture and Real Estate, 1981–1991* (College Station: Texas A&M University Press, 1993).

24. William Hauser, "The economic battle in America's office markets revisited: Dynamics of Class A, B, and C office space in the Chicago market" (ULI Research working paper 633, Urban Land Institute, Washington, DC, 1993).

25. Hunt and DeVries, *Planning Chicago.*

26. Author's calculation based on CoStar Property Analytics data for the six-county Chicago region, 2000. See also Robert Lang, *Office Sprawl: The Evolving Geography of Business* (Washington, DC: Brookings Institution Press, October 2000).

27. Author's calculation based on "Where Workers Work," Illinois Department of Employment Security, data for 2000, http://www.ides.illinois.gov/LMI/Pages/Where_Workers_Work .aspx. Most of these workers were employed in the advanced producer and corporate services (legal, accounting, communications) and finance and insurance sectors, and they worked in "office support occupations" as secretaries, administrative support, customer service representatives, and clerks.

28. Sofia Dermisi, "The impact of terrorism fears on downtown real estate Chicago office market cycles," *Journal of Real Estate Portfolio Management* 13, no. 1 (2007): 57–73.

29. John Handley, "Kinzie Station switches to rental: Developer sees overbuilding in condo market," *Chicago Tribune*, November 16, 2001.

30. Business Cycle Dating Committee, National Bureau of Economic Research, July 17, 2003, http://www.nber.org/cycles/july2003.html.

31. H. Lee Murphy, "City of big shoulders craves big office tower," *National Real Estate Investor* 49, no. 10 (October 1, 2007): 67.

32. Different estimates of available supply exist. These data were derived from CoStar, a private, subscription-based data service that covers buildings of all classes, sizes, and ownership (public or private, owner-occupied or multi-tenant). Because CoStar's records of buildings with non-office uses tend to be less exhaustive, I supplemented them with data from other sources, including the Goodman Williams Group database of downtown development. Thanks to Christine Williams for providing this information.

33. This figure does not take into account the buildings that were demolished or converted after 2000.

34. Bennett, *Third City*, 102.

35. Thanks to Sharon Haar for pointing this out.

36. Developers C and D.

37. Quoted in Susan Diesenhouse, "Architects' pace rapidly builds," *Chicago Tribune*, January 22, 2006.

38. Mid-rises such as Museum Park, Prairie District Homes, and Prairie House opened nearby. Deborah Johnson, "The new face of Chicago's South Loop," *Urban Land*, April 2002.

39. Author's calculation based on U.S. Bureau of the Census, Summary File 1, data for 2000 and 2010, http://www2.census.gov/census_2000/datasets/Summary_File_1/Illinois/.

40. Prescient for their time, Mary Ludgin and Louis Masotti in 1985 heralded the emergence of a "Super-Loop" that stretched from North Avenue to Roosevelt Road, and from Lake Michigan west to Racine Avenue. Ludgin and Masotti, "Downtown development, Chicago 1985–1986" (Chicago: Center for Urban Affairs and Policy Research, Northwestern University, 1986).

41. Author's calculation based on CoStar Property Analytics data for 2008.

42. Charles S. Suchar, "The physical transformation of metropolitan Chicago: Chicago's Central Area," in *The New Chicago*, eds. John P. Koval, Larry Bennett, Michael I. J. Bennett, Fassil Demissie, Roberta Garner, and Kiljoong Kim (Philadelphia: Temple University Press, 2006), 56–76.

43. Dennis Rodkin, "Who will buy the condos?," *Chicago Magazine*, May 9, 2008.

44. Chicago Loop Alliance, *2011 Loop Economic Study & Impact Report*, 2011, http://loopchicago.com/_files/2011_Loop_Economic_Study_FINAL.pdf. The mixed-use University Center project, for example, represented a unique collaboration of three colleges undergoing downtown expansions. The dormitories also took advantage of earlier conversions of office buildings to classroom space, such as DePaul's repurposing of the Goldblatt's Department Store building on State Street and Robert Morris's conversion of the Sears Building on Congress.

45. The National Bureau of Economic Research dates the Great Recession as starting in December 2007 and lasting through June 2009. Business Cycle Dating Committee, accessed January 20, 2015, http://www.nber.org/cycles/cyclesmain.html.

46. Lender B.

47. Colin Lizieri, "Occupier requirements in commercial real estate markets," *Urban Studies* 40, no. 5–6 (2003): 1151–69, 1152.

48. Chicago Metropolis 2020, *The 2002 Metropolis Index* (Chicago, 2001), http://www.metropolisstrategies.org/documents/02_intro.pdf.

49. Christopher Muellerleile, "Financialization takes off at Boeing," *Journal of Economic Geography* 9 (2009): 663–77.

50. Thanks to World Business Chicago for data on headquarters relocations to and from Chicago.

51. Author's calculation based on data from "Where Workers Work," 1998–2009.

52. Author's calculation based on "Quarterly Census of Employment and Wages," data from 2000–2005, U.S. Bureau of Labor Statistics, http://www.bls.gov/cew/data.htm.

53. John Slania, "Market learned from mistakes," *Crain's Chicago Business*, March 25, 2001.

54. I intentionally use 2008 data for jobs to capture employment trends before the worst of the 2007–2009 recession hit. These data are similar to those used by Miles to create his Space Market Index, one of the few measures of overbuilding. He divides the moving average of changes in employment by the moving average of changes in stock. Mike Miles, "A foundation for the strategic real estate allocation: The space market index," *Real Estate Finance* 14 (1997): 3–23.

55. Paul Krugman, "Boiling the frog," *New York Times*, July 13, 2009; see also Stanley Aronowitz, *Just around the Corner: The Paradox of the Jobless Recovery* (Philadelphia: Temple University Press, 2005).

56. Author's calculation based on "Local Area Unemployment Statistics," data from 2000–2010, U.S. Bureau of Labor Statistics, http://www.bls.gov/lau/.

57. John Burns and John McDonald, *Who Are Your Future Tenants? Office Employment in the United States, 2004–2014*, National Association of Realtors, January 2007,. http://www.rcasenc.com/documents/employment04-14.pdf; Todd Clark and Taisuke Nakata, "The trend growth rate of employment: Past, present, and future," *Economic Review* (Federal Reserve Bank of Kansas City) 91, no. 1 (2006): 43.

58. Lizieri, "Occupier requirements"; PricewaterhouseCoopers LLP and the Urban Land Institute, Emerging Trends in Real Estate, 2006, http://www.sddt.com/images/news/2006/02/10/EmergTrends-06.pdf; Norm Miller, "Downsizing and workplace trends in the office market," *CRE Real Estate Issues* 38, no. 3 (2013), http://www.cre.org/memberdata/pdfs/Downsizing_Workplace_Trends.pdf.

59. Some such firms consume up to 800 square feet per lawyer, compared with an average of 225 to 280 per employee in other sectors. Burns and MacDonald, *Who Are Your Future Tenants?*; Ryan Ori, "Law firms' newest case: Cut real estate," *Crain's Chicago Business*, October 28, 2013.

60. This was significantly higher than the 8.4 percent vacancy rate in rival markets such as midtown Manhattan. However, it was also lower than the vacancy rate in the suburban Chicago office market, which hit 23 percent at this time. See Terri Pristin, "Lower Manhattan: A relative bargain but filling up fast," *New York Times*, February 28, 2007.

61. Mark Gallagher and Antony P. Wood, "Fear of overbuilding in the office sector: How real is the risk and can we predict it?," *Journal of Real Estate Research* 17, no. 1 (1999): 3–32.

62. Author's calculation based on data from "Where Workers Work," 1998–2008.

63. Chicago Loop Alliance, *2011 Loop Economic Study & Impact Report.*

64. I use the per-employee space estimate of 280 square feet from CoStar's 2006 tenant survey, which includes leased common areas such as conference and storage rooms. *The CoStar Office Report* (CoStar Group Inc., 2006). See also Burns and McDonald, *Who Are Your Future Tenants?* This estimate, however, is conservative; researchers typically estimate space usage figures ranging from 150 to 210 square feet of net rentable area for all categories of office space combined. See, for example, Joseph Rabianski and Karen M. Gibler, "Office market demand analysis and estimation techniques," *Journal of Real Estate Literature* 15, no. 1 (2006): 37–56; Na-

than Schloss, "Use of employment data to estimate office space demand," *Monthly Labor Review* 107 (1984): 40–44. Note also that *new* employees are likely to occupy a different amount of space per capita than the overall industry average.

65. The estimate of 154 million square feet of office space in the downtown submarkets in 2009 was derived from CoStar Property Analytics data. See above and n. 32 in this chapter. It does not include office space available in other types of buildings such as retail establishments and hotels. The actual amount of available office stock, therefore, would be higher, as would be the estimate of surplus space.

66. Mark Heschmeyer, "Has the office vacancy rate become irrelevant?," *CoStar Group Real Estate Information, National News*, April 15, 2009, http://www.costar.com/News/Article/Has -the-Office-Vacancy-Rate-Become-Irrelevant-/111445.

67. These aggregate vacancy and absorption figures do not account for the fact that the downtown market is home to a variety of submarkets differentiated by age, quality, and location, a fact discussed in more depth in chapter 6.

68. Liz Ptacek, Nicholas Buss, and Robert Bach, *Myth or Reality: No Worse, No Better, No Hope? Catching Up with the Office Market* (Chicago: PNC Real Estate Finance and Grubb & Ellis, September 2003), 16.

69. Based on an analysis of address changes of Chicago region businesses derived from Dun & Bradstreet establishment data. See Rachel Weber, "Restless in place: The causes and planning implications of intra-metropolitan business relocations" (presented at the Association of Collegiate Schools of Planning Annual Meeting, Milwaukee, WI, October 2007).

70. No industrial uses appeared in the Central Area during the boom, although several warehouses were constructed on the fringes of the CBD in places such as the Fulton-Carroll Market District.

71. Author's calculation based on data from "Where Workers Work,"1998–2008. See table 4.2.

72. Population growth was even greater within the smaller geography of the Loop, increasing by a whopping 187 percent during this decade. Chicago Loop Alliance, *2011 Loop Economic Study & Impact Report.*

73. Eugenie Birch, *Who Lives Downtown Today (And Are They Any Different from Downtowners of Thirty Years Ago)?* (Cambridge, MA: Lincoln Institute of Land Policy, 2005), 3.

74. Chicago Loop Alliance, *2011 Loop Economic Study & Impact Report.*

75. Choose Chicago, *Market Analytics*, 2009, http://www.choosechicago.com/includes/ content/docs/media/Chicago-Detailed-Visitation-Report-2013-6-30.pdf.

76. Dennis Judd, "Constructing the tourist bubble," in *The Tourist City*, eds. Dennis Judd and Susan Fainstein (New Haven, CT: Yale University Press, 1999), 35–53.

77. Timothy Gilfoyle, *Millennium Park: Creating a Chicago Landmark* (Chicago: University of Chicago Press, 2006).

78. Goodman Williams Group and URS, *Millennium Park Economic Impact Study* (2005), http://www.goodmanwilliamsgroup.com/Pages/ProjectClientPages/Millennium_Park _Economic_Impact_Study.pdf.

79. Author's calculation based on U.S. Bureau of the Census, Summary File 1, data on households and housing units for 2000 and 2010, http://www2.census.gov/census_2000/datasets/ Summary_File_1/Illinois/ and http://www2.census.gov/census_2010/04-Summary_File_1/ Illinois/. New households can be formed when they relocate to the area from outside, when children move out of their parents' homes, or when related or unrelated individuals choose to live singly after previously sharing a residence.

80. Alby Gallun, "Condominium sales fall to a 7-year low—Unsold downtown units pile up as building boom continues," *Crain's Chicago Business*, November 10, 2008.

81. Alby Gallun, "Downtown dead zone," *Crain's Chicago Business*, May 12, 2008.

82. William Frey, *Metropolitan Magnets for International and Domestic Migrants* (Washington, DC: Brookings Institution, 2003).

83. Author's calculation based on "Community Population Survey," U.S. Bureau of Labor Statistics, 1998–2008, data downloaded from Integrated Public Use Microdata Series, http://cps .ipums.org/cps.

84. Ibid.

85. Deborah Myerson, "Converting the Historic Palmolive Building into Luxury Condominiums," in *Developing Condominiums: Successful Strategies* (Washington, DC: Urban Land Institute, 2006), 82–93.

86. Institute for Housing Studies, *Overview of the Chicago Housing Market* (Chicago: DePaul University, 2013), http://www.housingstudies.org/media/filer_public/2013/10/01/ihs_2013 _overview_of_chicago_housing_market.pdf.

87. Author's calculation based on "Community Population Survey," U.S. Bureau of Labor Statistics, 2002–2008, data downloaded from Integrated Public Use Microdata Series, http://cps .ipums.org/cps.

88. Alan Ehrenhalt, *The Great Inversion and the Future of the American City* (New York: Random House, 2012).

89. Giovanni Dosi, "Technological paradigms and technological trajectories: A suggested interpretation of the determinants and directions of technical change," *Research Policy* 11, no. 3 (1982): 147–62. For a discussion of these interactions in the property sector, see David Gann, *Building Innovation: Complex Constructs in a Changing World* (London: Thomas Telford Publishing, 2000).

90. Thanks to Michael Kaplan for pointing this out.

91. Broker F.

92. Quoted in Robert, "Lawyers and real estate experts talk office space."

93. Tenant A.

94. Facilities Manager A.

95. Productivity is defined as output per unit of labor, capital, or other measurable input. In theory, relocation should occur only if rental or operating cost savings combined with revenue increases from enhanced productivity exceed moving costs.

96. The five-year average lease rate per square foot for Class A office buildings constructed between 1998 and 2008 was $36.42, while for Class A buildings constructed between 1975 and 1991 it was $32.03. Author's calculation based on CoStar Property Analytics data, 2008–2013.

97. Sofia Dermisi, "Effect of LEED ratings and levels on office property assessed and market values," *Journal of Sustainable Real Estate* 1 (2009): 23–47; Piet Eichholtz, Nils Kok, and John M. Quigley, "The economics of green building," *Review of Economics and Statistics* 95, no. 1 (2013): 50–63; Jonathan Wiley, Justin Benefield, and Ken Johnson, "Green design and the market for commercial office space," *Journal of Real Estate Finance and Economics* 41, no. 2 (2010): 228–43.

98. Broker B. After downsizing and merging, many law firms still had to pay for vacant space (one estimate was that over 16 percent of all Chicago law firms' leased space was vacant) or sublease it at rates 15 to 40 percent below what they paid their landlords. Ori, "Law firms' newest case."

99. Matthew Kahn, Nils Kok, and John M. Quigley, "Commercial building electricity con-

sumption dynamics: The role of structure quality, human capital, and contract incentives" (working paper 18781, National Bureau of Economic Research, 2013), http://www.nber.org/papers/w18781. More problematically, some studies have found no evidence that LEED certification lowers either site or source energy consumption. Site energy consumption is the amount of heat and electricity consumed by a building as reflected in its utility bill. Source energy consumption represents the total amount of raw fuel that is required to operate the building and incorporates all transmission, delivery, and production losses. John Scofield, "Do LEED-certified buildings save energy? Not really," *Energy and Buildings* 41 (2009) 1386–90; John Scofield, "Efficacy of LEED-certification in reducing energy consumption and greenhouse gas emission for large New York City office buildings," *Energy and Buildings* 67 (2013): 517–24.

100. Facilities Manager A.

101. Author's calculation based on BOMA Experience Exchange Reports, "Chicago Downtown," 1998–2010, http://www.boma.org/research/Pages/eer.aspx. Expenses include utilities, maintenance of heating and air conditioning systems, and upkeep of the common areas, in addition to insurance and property taxes.

102. Eichholtz, Kok, and Quigley, "Economics of green building"; Sofia Dermisi and John McDonald, "Selling prices/sq. ft. of office buildings in Downtown Chicago: How much is it worth to be an old but Class A building?," *Journal of Real Estate Research* 32, no. 1 (2010): 1–21.

103. Michael Harper, Bhavani Khandrika, Randal Kinoshita, and Steven Rosenthal, "Non-manufacturing industry contributions to multi-factor productivity, 1987–2006," *Monthly Labor Review*, June 2010, 16–31.

104. Ameet Sachdev, "Recession hits top law firms' profits," *Chicago Tribune*, April 30, 2009; Steven Strahler, "Economy still punishing big Chicago law firms," *Crain's Chicago Business*, February 28, 2013.

105. Thomas Corfman, "Kirkland firm joins flight to new office towers," *Chicago Tribune*, June 10, 2005.

106. Author's calculation based on IBIS World Industry Statistics (law firms), 2002–2009, http://clients1.ibisworld.com/; see also Ori, "Law firms' newest case," reporting on a Cushman & Wakefield survey. That rents ate up a larger percentage of law firm revenues could be due to higher rents, declining revenues, or both.

107. For instance, law firm Keck, Mahin & Cate entered into a long-term lease for new office space at 77 West Wacker Drive even though it subsequently reported difficulties paying the rent. Amanda Robert, "Picking up the pieces after a law firm collapses," *Chicago Lawyer*, June 1, 2011.

108. Bennett, *Third City*, 195.

109. Susan Diesenhouse, "Despite sluggish leasing, buildings rise," *Chicago Tribune*, December 31, 2006.

110. Chicago Loop Alliance, *2011 Loop Economic Study & Impact Report*.

Chapter Five

1. Alby Gallun, "Zell's fancy turns away from real estate," *Crain's Chicago Business*, March 21, 2005.

2. EOP bought out fellow REITs Spieker (with a base in Silicon Valley), Beacon Properties (with holdings in Boston and other parts of the Northeast), and Cornerstone Properties Trust (owned primarily by a Netherlands-based pension fund) as well as other investment partnerships in Chicago. Steve Daniels, "Vulture no more," *Crain's Chicago Business*, May 7, 2001.

3. Alby Gallun, "New Equity Office executive leading reorganization finale," *Crain's Chicago Business,* November 17, 2003.

4. Daniels, "Vulture no more."

5. Ibid.

6. Alby Gallun, "Zell eyes towering cash-out: Financier's REIT hawking stakes in Chicago Title site, other top holdings," *Crain's Chicago Business,* September 22, 2003.

7. Eddie Baeb and Gregory Meyer, "Zell pays off for the patient," *Crain's Chicago Business,* February 12, 2007.

8. David Roeder, "Equity Office sale a winner for Zell and Blackstone," *Chicago Sun-Times,* November 26, 2006.

9. EOP's Chicago holdings were split up by location to boost the price. GE Real Estate paid $1.05 billion for nine suburban properties, including Oakbrook Terrace Tower. A month later Tishman Speyer paid $1.72 billion for seven downtown buildings, including the Civic Opera Building and the dual towers flanking the Chicago Mercantile Exchange.

10. Reporter B.

11. Susan Diesenhouse, "Despite sluggish leasing, buildings rise," *Chicago Tribune,* December 31, 2006.

12. Karin Knorr Cetina and Urs Bruegger, "Global microstructures: The virtual societies of financial markets," *American Journal of Sociology* 107, no. 4 (2002): 905–50; Saskia Sassen, "Chicago's deep economic history: Its specialized advantage in the global network," in *Chicago's Geographies: Metropolis for the 21st Century,* ed. Richard Greene, Mark Bouman, and Dennis Grammenos (Washington, DC: Association of American Geographers, 2006), 75–86.

13. Colin Lizieri, *Towers of Capital: Office Markets & International Financial Services* (Oxford: Wiley-Blackwell, 2009); Dariusz Wójcik, "Financial centre bias in primary equity markets," *Cambridge Journal of Regions, Economy and Society* 2, no. 2 (2009): 193–209; Dariusz Wójcik, "Securitization and its footprint: The rise of the US securities industry centres 1998–2007," *Journal of Economic Geography* 11, no. 6 (2011): 925–47.

14. Mortgage Broker A.

15. Author's calculation based on the annual 10-K filings from the twelve largest publicly traded office REITs on *Forbes* magazine's REIT Gold List in 2007. Only 6 percent (2003), 10 percent (2005), and 2 percent (2007) of the total value of their collective holdings was from Chicago buildings. The REIT Gold List tracks those publicly traded REITs with the largest holdings in the United States. Dorothy Pomeranz, "The REIT Gold List," *Forbes,* accessed January 20, 2015, http://www.forbes.com/2007/02/05/reits-real-estate-tinvestment-pfcz_kb_0208reits_table .html.

16. Author's calculation based on an average of annual averages from CoStar Property Analytics data, 2001–2007. The difference is partly attributable to the relative sizes of the markets: Manhattan hosts more than twice the rentable office space Chicago does, and, on average, eighteen more major sales transactions take place per year there than in Chicago. But even comparable office markets outsold Chicago: e.g., office buildings in Washington, DC, sold for $230 more per square foot.

17. "NCREIF Property Index Returns," 2008, National Council of Real Estate Investment Fiduciaries, https://www.ncreif.org/property-index-returns.aspx.

18. Author's calculation based on CoStar Property Analytics data, 2000–2010.

19. Sofia Dermisi and Jon DeVries, *Economic Impact Study of Chicago Downtown Office Market* (Chicago: Building Owners and Managers Association, 2006). See also Lorene Yue, "Chicago

tops in commercial property tax rates: BOMA study," *Crain's Chicago Business,* June 8, 2006. The prevalence of exemptions, deductions, abatements, and appeals probably brings Chicago's property taxes more in line with those of other cities. Even if taxes in Chicago are not higher, they are probably more confusing, given the multiple forms of relief available to property owners.

20. Ross Miller, *Here's the Deal: The Buying and Selling of a Great American City* (New York: Knopf, 1996), 184.

21. Quoted in Diesenhouse, "Despite sluggish leasing, buildings rise."

22. Steve Kerch, "New tenants difficult to find for downtown office space, brokers say," *Chicago Tribune,* March 4, 1990.

23. Edward Glaeser, Joseph Gyourko, and Albert Saiz, "Housing supply and housing bubbles," *Journal of Urban Economics* 64, no. 2 (2008): 198–217.

24. While the downtown hosted approximately 40 vacant and developable parcels suitable for office uses in 2000, no vacant parcels appropriate for office construction in Washington, DC, or Manhattan were available as of 2010. Developer D.

25. Micah Maidenberg, "Real estate employment falls in Chicago area," *Crain's Chicago Business,* September 24, 2012. These data include brokers, property managers, lawyers, developers, and real estate financiers as well as construction workers.

26. Author's calculation based on "Current Employment Statistics," 2000–2010, U.S. Bureau of Labor Statistics, http://www.bls.gov/ces/.

27. Chicago is not an exception in this regard. Andrew Wood, "The scalar transformation of the U.S. commercial property development industry," *Economic Geography* 80, no. 2 (2004): 119–40.

28. Miles Berger, *They Built Chicago: Entrepreneurs Who Shaped a Great City's Architecture* (Chicago: Bonus Books, 1992).

29. Mortgage Broker A.

30. Miller, *Here's the Deal;* Milton Rakove, *Don't Make No Waves—Don't Back No Losers: An Insider's Analysis of the Daley Machine* (Bloomington: Indiana University Press, 1976); William Grimshaw, *Bitter Fruit: Black Politics and the Chicago Machine, 1931–1991* (Chicago: University of Chicago Press, 1995).

31. Author's calculation based on CoStar Property Analytics data, 2010.

32. Author's calculation based on CoStar Property Analytics data, 2001–2007.

33. Asset Manager A.

34. Asset Manager A and Broker C.

35. Larry Bennett, *The Third City: Chicago and American Urbanism* (Chicago: University of Chicago Press, 2010).

36. Robert DeYoung and Thomas Klier, "Why Bank One left Chicago: One piece in a bigger puzzle," *Chicago Fed Letter* no. 201, April 2004, https://www.chicagofed.org/publications/chicago-fed-letter/2004/april-201; Tara Rice and Erin Davis, "The branch banking boom in Illinois: A byproduct of restrictive branching laws," *Chicago Fed Letter* no. 238, May 2007, https://www.chicagofed.org/publications/chicago-fed-letter/2007/may-238; Gary Dymski, *The Bank Merger Wave* (New York: M. E. Sharpe, 1999).

37. Ted Pincus, "Minimal bureaucracy works wonders for MB," *Chicago Sun-Times,* April 8, 2008.

38. With this historical legacy, it is no wonder that Chicago was ground zero for both discriminatory lending practices and the grassroots activism challenging them. See Gregory Squires, *Redlining to Reinvestment: Community Responses to Urban Disinvestment* (Philadelphia:

Temple University Press, 1992); Alex Schwartz, "From confrontation to collaboration? Banks, community groups, and the implementation of community reinvestment agreements," *Housing Policy Debate* 9, no. 3 (1998): 631–62.

39. Mortgage Broker A.

40. Richard Herring, *The Collapse of Continental Illinois National Bank and Trust Company: The Implications for Risk Management and Regulation* (Philadelphia: Wharton Financial Institutions Center, 1994), http://fic.wharton.upenn.edu/fic/case%20studies/continental%20full.pdf.

41. Steve Daniels, "LaSalle Bank's future looks murky," *Crain's Chicago Business*, March 26, 2007.

42. Steve Daniels, "Only game in town," *Crain's Chicago Business*, July 25, 2005.

43. Mortgage Broker B.

44. Steve Daniels, "LaSalle loans bite B of A," *Crain's Chicago Business*, November 24, 2008.

45. Nick Timiraos and Jessica Holzer, "Corus Bank is the latest to be seized by regulators," *Wall Street Journal*, September 12, 2009.

46. Because of their larger consumer operations and stronger reserves, these banks each had less than 10 percent of their loans in commercial real estate projects (in 2009).

47. The Bank of Montreal bought Harris Bank in 1984, Banc One bought First National in 1998, Minneapolis-based First Bank bought Boulevard in 1994, and JPMorgan Chase purchased Bank One in 2004 (which had purchased Chicago-based First Chicago NBD in 1998).

48. Broker C.

49. In the 2000s insurance companies such as Manulife (Canada) and SEB (Sweden) and merchant banks such as Bankhaus Metzler (Germany), HypoVereinsbank (Germany), Helaba (Germany), and the Anglo-Irish Bank entered the Chicago market. Susan Diesenhouse, "Foreign capital increases in Chicago," *Chicago Tribune*, March 14, 2007. More than 10 percent of Chicago's office buildings were owned by foreign investors in 1985, with Japanese banks leading the way. See John Oharenko and Anatole Kolomayets, *The 1985 Real Estate Capital Market Report* (Chicago: Baird & Warner, 1985).

50. David Roeder, "Builders up against tighter lending rules," *Chicago Sun-Times*, October 23, 2000.

51. Daniels, "Vulture no more."

52. Quoted in John Slania, "Market learned from mistakes," *Crain's Chicago Business*, March 25, 2001.

53. Roeder, "Builders up against tighter lending rules."

54. Appraiser A.

55. Quoted in Thomas Corfman, "USG HQ sold for $168 mil," *Crain's Chicago Business*, November 18, 2006.

56. Alan Lev of the Belgravia Group, "Downtown Residential: Poised for a Rebound" (Urban Land Institute presentation, Chicago, October 28, 2010). An example of such thinking was developer Teng and Associates, which began building the Shangri La Hotel and Waterview Tower *before* it received construction financing (the project eventually stalled out).

57. Future condo owners played games as well, e.g., hiding the fact that they lacked sufficient income to make debt service payments.

58. Author's calculation based on CoStar Property Analytics data, 1998–2009. Some of these were portfolio sales, in which more than one property was sold as part of a package.

59. In the entire metro region, sales totaled almost $11.9 billion in 2007. Acquisitions of office buildings constituted the bulk of CRE trading volumes in the region; trailing far behind, but

making up the majority of individual transactions, was retail, followed by apartment buildings. Thomas Corfman, "'07 was booming, until the big bust," *Crain's Chicago Business*, January 21, 2008.

60. However, for the reasons laid out earlier, many Chicago-based private equity firms were slow to invest in Chicago companies and pursued opportunities elsewhere. Just 12 percent of the financial transactions reported by Chicago private equity firms in 2004 involved local businesses. Sarah Klein, "The out-of-towners," *Crain's Chicago Business*, February 21, 2005.

61. The most active brokers at the acquisitions peak in 2006 were CB Richard Ellis, Eastdil, Colliers Bennett and Kahnweiler, and Jones Lang LaSalle. *Crain's Chicago Business*, "Crain's List: Chicago's largest commercial real estate brokers by transaction volume," December 4, 2006.

62. Jeffrey Bramson and Jaime Fink, Holliday Fenoglio Fowler LP, interview, *Crain's Chicago Business*, December 3, 2007.

63. Despite their historic heights, these prices were still considered puny by coastal standards: some office buildings in Manhattan, for example, were being priced at over $1,000 per square foot. Catherine Curan, "Office sales breaking the bank in Manhattan," *The RealDeal*, October 26, 2007, http://therealdeal.com/issues_articles/office-sales-breaking-the-bank-in -manhattan/.

64. Broker C. The capitalization rate for 190 South LaSalle was less than 1 percent when it was sold in 2006.

65. Thomas Corfman, "Chicago-area office sales to set record," *Crain's Chicago Business*, December 17, 2007. In the case of Prudential Plaza, tenants had moved to newer skyscrapers. In the case of the John Hancock Center, tenants went bankrupt or were acquired in what could have been read as a sign of imminent crisis.

66. Quoted in Diesenhouse, "Despite sluggish leasing, buildings rise."

67. Ibid.

68. Sofia Dermisi and John McDonald, "Selling prices/square foot of office buildings in downtown Chicago: How much is it worth to be an old but Class A building?," *Journal of Real Estate Research* 32, no. 1 (2010): 1–22.

69. Quoted in Baeb and Meyer, "Zell pays off for the patient."

70. Asset Manager A.

71. Peter Colwell, Henry Munneke, and Joseph Trefzger, "Chicago's office market: Price indices, location and time," *Real Estate Economics* 26, no. 1 (1998): 83–106.

72. Appraiser A.

73. Author's calculation based on FTSE All REITs Index, total debt, equity and market capitalization data, 2008, National Association of Real Estate Investment Trusts, https://www.reit .com/investing/industry-data-research/industry-data.

74. William Cronon, *Nature's Metropolis: Chicago and the Great West* (New York: Norton, 1992).

75. *The Economist* 1986, cited in Chris Muellerleile, "Turning financial markets inside out: Polanyi, performativity and disembeddedness," *Environment and Planning A* 45 (2013): 1625–42. The trading of these instruments thrived until over-the-counter trades began to cut out the exchanges in the mid-1990s. Still, Chicago sustained its reputation as a global financial center through the Millennial Boom (e.g., Eurex, the world's largest derivatives exchange, located its all-electronic derivatives exchange in the city in 2004).

76. Alan Rabinowitz, *The Real Estate Gamble* (New York: American Management Association, 1980).

77. Nelson Peach, *The Security Affiliates of National Banks* (Baltimore: Johns Hopkins Press, 1941).

78. Herring, *Collapse of Continental Illinois National Bank and Trust Company.*

79. Jason Kravitt, *Securitization of Financial Assets*, 3rd ed. (New York: Aspen / Wolters Kluwer, 2013).

80. Dariusz Wójcik, "The dark side of NY–LON: Financial centres and the global financial crisis," *Urban Studies* 50, no. 13 (2013): 2736–52. Although Chicagoans took umbrage at being, once again, forced into second place, they also experienced a kind of schadenfreude in 2008 when all fingers pointed to Wall Street as the main production point of the crisis. Lender C.

81. Lender C.

82. Mortgage Broker A.

83. Lender B. Another informant (Lender G) noted that LaSalle's appeal lay in the fact that it "did not act like a Wall Street bank. It had a much more personal style."

84. Lender G.

85. Maura Webber Sadovi, "Back to basics for Tishman," *Wall Street Journal*, December 22, 2010. In the case of 353 North Clark, for example, the building's construction loan had not yet been paid off, so when Tishman Speyer bought it, the CMBS loan acted as a kind of takeout financing. See Alby Gallun, "Market revives for local CMBS loans," *Crain's Chicago Business*, June 27, 2011.

86. Mortgage Broker B.

87. Ibid.

88. For a similar discussion of clustering in London's MBS industry, see Michael Pryke and Roger Lee, "Place your bets: Towards an understanding of globalisation, socio-financial engineering and competition within a financial centre," *Urban Studies* 32, no. 2 (1995): 329–44.

89. Steve Bergsman, "Sweet synergies: JPMorgan's CMBS business shot up the issuer rankings following merger with Bank One," *Mortgage Banking* 66, no. 4 (2006): 46–49.

90. Lender E.

91. Ibid.

92. Quoted in Corfman, "'07 was booming."

93. John Logan and Harvey Molotch, *Urban Fortunes: The Political Economy of Place* (Berkeley: University of California Press, 1987); Bernard Frieden and Lynn Sagalyn, *Downtown, Inc.: How America Rebuilds Cities* (Cambridge, MA: MIT Press, 1989).

94. Other city council members deferred to the Loop's representatives when it came to development decisions downtown, granting what is called "aldermanic privilege." Dick Simpson, *Rogues, Rebels, and Rubber Stamps: The Politics of the Chicago City Council from 1863 to the Present* (Boulder, CO: Westview Press, 2001).

95. These groups included the Central Loop Alliance, Friends of Downtown, the Building Owners and Managers Association, the Metropolitan Planning Council, and the Chicago Central Area Committee.

96. Bennett, *Third City*; D. Bradford Hunt and Jon DeVries, *Planning Chicago* (Chicago: APA Planners Press, 2013); Keith Koeneman, *First Son: The Biography of Richard M. Daley* (Chicago: University of Chicago Press, 2013).

97. Planner G.

98. Hunt and DeVries, *Planning Chicago.* Particular real estate firms were part of Daley's trusted inner circle, including bond underwriter William Blair and Company, which sold public

debt instruments to global investors, and architecture firm Skidmore Owings Merrill (SOM), which wrote the plans of 1973, 1983, and 2003 and for the 2016 Olympics bid.

99. Miller, *Here's the Deal*, 184.

100. Lois Wille, *At Home in the Loop* (Carbondale: Southern Illinois University Press, 1998), 1.

101. Larry Bennett, Janet Smith, and Patricia Wright, *Where Are Poor People to Live? Transforming Public Housing Communities* (New York: M. E. Sharpe, 2006).

102. James Warren, "My kind of technocracy," *New York Times*, August 14, 2010.

103. The Michigan/Cermak TIF was the first district in the South Loop, followed by the Near South TIF.

104. City of Chicago Annual Financial Analysis 2011, July 29, 2011, 7, https://ia601003.us .archive.org/31/items/228547-city-of-chicago-annual-financial-analysis-2011/228547-city-of -chicago-annual-financial-analysis-2011.pdf.

105. This figure does not include payments made to other taxing jurisdictions (such as the Chicago Transit Authority) that are governed through intergovernmental agreements. It also does not include projects for which land sales occurred before the boom, even if construction took place during it.

106. Christine Raguso, acting commissioner, Chicago Department of Community Development (Lambda Alpha International—Ely Chapter presentation, Chicago, April 15, 2009).

107. In the Central Area the city used TIF for private debt financing (39 percent), public works and improvements (32 percent), and property assembly and site preparation (18 percent). *Central Area Action Plan*, August 20, 2009, http://www.cityofchicago.org/dam/city/depts/zlup/ Planning_and_Policy/Publications/Central_Area_Action_Plan_DRAFT/CAAPRevisedIntro .pdf.

108. Deborah Johnson, "The new face of Chicago's South Loop," *Urban Land*, April 2002.

109. Ibid.

110. Planner A.

111. These were the Near West (1996), River South (1997), Canal Congress (1998), and River West (2001) TIF districts. Subsidized developers included Hines, which received assistance with its ABN Amro Tower; Fifield, which received TIF funds for its eighteen-story 550 West Adams; and Buck, for its buildup of 550 West Jackson.

112. Timothy Gilfoyle, *Millennium Park: Creating a Chicago Landmark* (Chicago: University of Chicago Press, 2006).

113. Rachel Weber, "Selling city futures: The financialization of urban redevelopment policy," *Economic Geography* 86, no. 3 (2010): 251–74.

114. Developer notes were used in approximately 70 of the 171 TIF deals between 1997 and 2007.

115. The city often issued developer notes as two discrete products. The first type of note was a commitment to paying a certain amount of principal and interest from the property tax increment upon completion of the development project. These notes were typically purchased by institutional investors. A second type was weighed down with locally specific, "public benefit" requirements related to design guidelines, job creation numbers, and quotas for women and minorities. These obligations were often purchased by local banks such as LaSalle and First Chicago. Weber, "Selling city futures."

116. Rebecca Hendrick, Martin Luby, and Jill Mason Terzakis, "The Great Recession's impact on the City of Chicago," *Municipal Finance Journal* 32 (Spring 2011): 33–69; Office of the Inspector General, City of Chicago, *Budget Options for the City of Chicago*, October 2010, http://

chicagoinspectorgeneral.org/wp-content/uploads/2011/02/IGO-Budget-Options-for-the
-City-of-Chicago-October-2010.pdf.

117. Civic Federation, *Financial Challenges for the New Mayor of Chicago: Analysis and Rec-ommendations*, February 14, 2011, www.civicfed.org/ . . . /Financial%20Challenges%20for%20the%20New%20Mayor.pdf. Its obligations are partly to blame for the city's deficit, which ballooned from $58.3 million in 2002 to $500 million in 2010. City of Chicago, *Annual Financial Analysis 2011*, https://ia601003.us.archive.org/31/items/228547-city-of-chicago-annual-financial-analysis-2011/228547-city-of-chicago-annual-financial-analysis-2011.pdf.

118. Hunt and DeVries, *Planning Chicago*.

119. City of Chicago, *Planning Principles for Chicago's Central Area* (Chicago: Department of Planning, September 1991). Thanks to Jim Peters for providing this document.

120. Joseph Schwieterman, Dana M. Caspall, and Jane Heron, *The Politics of Place: A History of Zoning in Chicago* (Chicago: Lake Claremont Press, 2006).

121. Ibid.

122. Planning scholar Melville Branch disparaged PDs for undermining the essence of comprehensive planning: "The ease with which waivers were granted to developers in return for amenity provisions makes nonsense of town planning." Branch, "Don't call it city planning: Misplaced densification in large US cities," *Cities* 3 (November 1986): 290–97. More ambivalent, former Chicago Commissioner of Planning David Mosena noted, "The bonus system is lax. We've probably become the easiest big city among the top ten in which to build something." Quoted in John McCarron, "Archaic zoning laws stifling downtown," *Chicago Tribune*, March 8, 1987.

123. Thanks to Ben Teresa for his analysis of PD agreements.

124. Blair Kamin, "High anxiety," *Chicago Tribune*, November 13, 2005.

125. Planner A.

126. Kamin, "High anxiety."

Chapter Six

1. The description of Village Green's redevelopment of 188 West Randolph is based on Alby Gallun, "188 W. Randolph going residential," *Crain's Chicago Business*, March 19, 2005, and "Developer seeks TIF district for Loop apartment project," *Crain's Chicago Business*, July 8, 2009.

2. Quoted in Gallun, "Developer seeks TIF district."

3. Lynn Becker, "Welfare queen," *Architecture Chicago Plus*, March 10, 2010. http://arcchicago.blogspot.com/2010/03/welfare-queen.html. Other perspectives on the issue associate such instances of public support for historic preservation with neoliberalism. See David Wilson, "Making historical preservation in Chicago: Discourse and spatiality in neo-liberal times," *Space and Polity* 8, no. 1 (2004): 43–59.

4. Quoted in John Slania, "Market learned from mistakes," *Crain's Chicago Business*, March 25, 2001.

5. Author's calculation based on CoStar Property Analytics data, 1997–2006. Leasing picked up in 1997, peaked in the first quarter of 2000, and hovered around 8 million square feet per quarter until the first quarter of 2003. It climbed upward after that, then started tapering off in 2006, reaching a low point in the first quarter of 2008, when only 2 million square feet of office space changed hands.

6. The average term for direct-lease Class A space (new deals) peaked at 10 years in late 2004

and fell to 7.9 years by 2009. MB Financial, *2011 Chicago Market Overview, Second Quarter 2011*, http://www.mbres.com/pdfs/research/mbre-cmo/mbre-cmo_2011q2.pdf.

7. Prices for sublet space were still too high for tenants of Class B and C space to afford. This situation contrasts with that in the early 1990s, when many Class A subleases were marketed with asking rents below those of direct-lease Class C space. William Hauser, "The economic battle in America's office markets revisited: Dynamics of Class A, B, and C office space in the Chicago market" (ULI Research working paper 633, Urban Land Institute, Washington, DC, 1993), 4.

8. Based on the twenty largest law firms in 2004. *Crain's Chicago Business*, "Crain's List: Chicago's largest law firms," November 1, 2004.

9. Based on geo-coded establishment-level data from Dun & Bradstreet, 1999–2003, for the five-county Chicago region. The zip codes that received the most establishments and those that sent the most establishments were either the same zip codes or were located adjacent to each other. Rachel Weber, "Restless in place: The causes and planning implications of intra-metropolitan business relocations" (presented at the Association of Collegiate Schools of Planning Annual Meeting, Milwaukee, WI, October 2007).

10. Borrowed from the title of Paulette Thomas, "Why did the law firm cross the street?: Houston developer lures Sidley to new skyscraper," *Crain's Chicago Business*, December 6, 2004.

11. Liz Ptacek, Nicholas Buss, and Robert Bach, *Myth or Reality: No Worse, No Better, No Hope? Catching Up with the Office Market* (Chicago: PNC Real Estate Finance and Grubb & Ellis, September 2003).

12. Even those law firms that renewed their leases rather than move reduced their space requirements to try to cut back on their real estate spending. Ameet Sachdev, "Lawyers losing a prized perk," *Chicago Tribune*, November 12, 2003.

13. See also Hauser, "Economic battle in America's office markets revisited."

14. Quoted in Amanda Robert, "Lawyers and real estate experts talk office space," *Chicago Lawyer*, March 4, 2011.

15. Grubb & Ellis, *2006 Real Estate Forecast: Great Lakes* (Northbrook, IL: Grubb & Ellis, 2005), 8.

16. Tenant A.

17. Tenant A and Facilities Manager A.

18. Thomas, "Why did the law firm cross the street?"

19. This self-cannibalizing behavior may be dictated by the buildings' own investors. Igal Charney, "Three dimensions of capital switching within the real estate sector: A Canadian case study," *International Journal of Urban and Regional Research* 25 (2001): 740–58.

20. Developer C.

21. See also Mark McCain, "Office brokerage fees; fuzzy rules and high stakes make a litigious climate," *New York Times*, September 3, 1989.

22. "Quarterly Census of Employment and Wages: Private Lessors of Nonresidential Buildings, NAICS 531120," 2001–2011, U.S. Bureau of Labor Statistics, http://www.bls.gov/cew/cewbultn09.htm.

23. Jones Lang LaSalle acquired Staubach in 2008.

24. Jeffrey Hornstein, *A Nation of Realtors: A Cultural History of the Twentieth-Century American Middle Class* (Durham, NC: Duke University Press, 2005).

25. Broker D.

26. Facilities Manager A.

27. Real Estate Consultant A.

28. Broker A. Several famous sports figures, such as former Chicago Bears tight end Desmond Clark, led or joined Chicago brokerages.

29. Commercial commissions are typically negotiated as part of each individual lease transaction and, in contrast to residential transactions, lack a standardized percentage. David Ibata, "Cash, cars, trips: Price of survival in office leasing market," *Chicago Tribune*, May 4, 1986. Typically a share of the total commission goes to the tenant's broker and the remainder goes to the listing broker representing the landlord. Individual agents then split their part of the commission with their brokerage company to cover overhead costs.

30. McCain, "Office brokerage fees."

31. Of landlords leasing Class A and B buildings in the Loop in 2008, 35 percent were offering bonus commissions above the average $1.25 per rentable square foot. M&J Wilkow, *Chicago Office Market Report 2009*, http://www.wilkow.com/site/epage/93150_804.htm.

32. Asset Manager A.

33. Broker G.

34. Office landlords were not the only ones offering concessions. Residential developer American Invsco offered to pay condo assessments and property taxes on its units for two or three years, and in some cases it promised to pay rent as well, guaranteeing investors a revenue stream to help defray monthly mortgage payments. Alby Gallun and Thomas Corfman, "Trouble in the towers," *Crain's Chicago Business*, June 16, 2008.

35. During the 1980s overbuilding allowed some tenants to pay what amounted to negative rents. Broker B.

36. In addition to these benefits, Chicago brokers could cap year-to year increases in those expenses that varied with occupancy (e.g., janitorial costs) and could limit what are called "gross up provisions" that would require tenants to pay more than their proportional share of these expenses if the building was not fully occupied.

37. M&J Wilkow, *Chicago Office Market Report 2009*.

38. MB Financial, *2012 Chicago Market Overview, Second Quarter 2012*, http://www.mbres.com/pdfs/research/mbre-cmo/mbre-cmo_2012q2.pdf. Tenant improvement allowances for Class A buildings between 2000 and 2009 averaged $40 per square foot but topped $70 in several cases.

39. Broker D.

40. Brokers A and C.

41. See Building Owners and Managers Association of Chicago, *BOMA/Chicago Economic Impact Study 2012*, April 9, 2013, 9, https://www.bomachicago.org/sites/default/files/2012_economic_impact_study_-_full_report_-_final_4-9-2013_0.pdf; Brokers A, B, and C. Broker A noted, "If you have to provide that information [on concessions], you can do it with tricks. For example, someone quotes you "face rents" for two years but by that time all the free rent has burned off and rents sound higher. Or they don't tell you about the tenant improvement allowance or the moving costs, or how the new owner paid off the old lease for a year. Information is only incompletely revealed and not directly observed."

42. Jim Costello, "The EOP effect," in *About Real Estate: A Free Torto Wheaton Publication* 8, no. 17, May 17, 2007.

43. Joe Gose, "Concession crazy," *National Real Estate Investor*, June 1, 2003.

44. However, as with the case of inter-metropolitan moves, such "but for" claims could not be verified, and the city had to negotiate these deals in the context of imperfect information,

good faith, and an imagined worst-case scenario of headquarters fleeing Chicago. Rachel Weber, "Do better contracts make better economic development incentives?," *Journal of the American Planning Association* 68, no. 1 (2002): 43–55.

45. Thanks to Ben Teresa for compiling data on these TIF-assisted relocations.

46. Thomas Corfman, "USG gets subsidy to remain in city," *Chicago Tribune*, April 7, 2004.

47. Thomas Corfman, "Amid iffy market, Fifield plans his 5th midrise in West Loop," *Chicago Tribune*, July 28, 2004.

48. Thomas Corfman, "USG HQ sold for $168 mil," *Crain's Chicago Business*, November 18, 2006.

49. Planner A.

50. Corfman, "USG gets subsidy to remain in city."

51. Ibid.

52. Elsewhere I characterize this process as one of "cannibalization." The difference in tone between these two analogous terms revolves around whether occupant upgrading occurs, whether one focuses on the occupant or the space occupied, and whether such processes take place in growing or stagnant markets. See Homer Hoyt, *One Hundred Years of Land Values in Chicago: The Relationship of the Growth of Chicago to the Rise of Its Land Values, 1830–1933* (Chicago: University of Chicago Press, 1933); William Baer and Christopher Williamson, "The filtering of households and housing units," *Journal of Planning Literature* 3 (Spring 1988): 127–52; Richard Harris, "'Ragged urchins play on marquetry floors': The discourse of filtering is reconstructed, 1920s–1950s," *Housing Policy Debate* 22, no. 3 (2012): 463–82.

53. See Figure 4.11. Vacancy rates narrowed from a difference of 5.2 percentage points in 2003 to −1.3 in 2010, when Class B and C vacancy rates dropped below those of Class A buildings.

54. Ptacek, Buss, and Bach, *Myth or Reality*; MB Financial, *2011 Chicago Market Overview, Second Quarter 2011*.

55. Brokers B and D. Empirical studies of housing filtering have drawn similar conclusions. See, for example, Ray Forrest and Alan Murie, "The dynamics of the owner-occupied housing market in Southern England in the late 1980s: A study of new building and vacancy chains," *Regional Studies* 28, no. 3 (1994): 275–89.

56. Reporter B.

57. Appraisal Research Counselors, "Office market analysis, City Of Chicago," 2008, 5, www .appraisalresearch.com /download/files/100077.pdf.

58. Brokers B, F, and G; Facilities Manager A; Lenders B and C.

59. Quoted in Robert, "Lawyers and real estate experts talk office space."

60. Author's calculation based on CoStar COMPS data, 2008.

61. Sofia Dermisi and John McDonald, "Selling prices/square foot of office buildings in downtown Chicago: How much is it worth to be an old but Class A building?," *Journal of Real Estate Research*, 32, no. 1 (2010): 1–22.

62. Ibid. Dermisi and McDonald consider "trophy" buildings to be those Class A towers built before 1972 that are architecturally significant (because of their exterior facades or heights) and have been well maintained (e.g., the Rookery). Brokers often referred to these historic buildings as "Class A minus." Some of these buildings (Dermisi and MacDonald count ten of them) held onto their values, posting sale prices 90 percent higher than Class B buildings of the same age.

63. Author's calculation based on CoStar Property Analytics data, 2008.

64. Facilities Manager A.

65. Planner A.

66. Broker D and Broker G. Examples include the austere glass-and-steel 200 North La-Salle (1984), the postmodern granite-clad AT&T building at 227 West Monroe (1989), and the stepped-roofed 203 North LaSalle (1985).

67. The top ten floors of the Mid-Continental Plaza at 55 East Monroe (1972) were reclad and converted into apartments, but this modernist building was the exception rather than the rule.

68. Architect A; Broker H.

69. Fred A. Bernstein, "Frank Gehry (a part owner) helps develop a landmark," *New York Times*, November 16, 2010.

70. Thomas Corfman, "Zombie fears stalk Tishman in Loop," *Crain's Chicago Business*, December 7, 2009.

71. Thomas Corfman, "Office tower at 500 W. Monroe flirts with foreclosure—again," *Crain's Chicago Business*, August 23, 2010.

72. Most of the condo development delinquencies were located in the South Loop. Alby Gallun and David Lee Matthews, "The complete guide to Chicago's condo collapse," *Crain's Chicago Business*, October 8, 2012. See also Alby Gallun, "Foreclosure looms for South Loop's huge Central Station project," *Crain's Chicago Business*, October 17, 2011.

73. Author's calculation based on Cook County Recorder of Deeds mortgage data, compiled by Record Information Services, 2008–2011, http://www.public-record.com/content/databases/foreclosures/index.asp.

74. Alby Gallun, "Banks keep chipping away at bad real estate debt," *Crain's Chicago Business*, November 25, 2013.

75. CMBS-funded assets considered "distressed"—whose future ownership and capital structure were in question—included One Prudential Plaza (1955), 200 North LaSalle (1984), 180 North LaSalle (1972), 200 South Wacker (1981), and 111 West Jackson (1961). This is not an exhaustive list, as properties are not recorded as "distressed" when lenders extend loan terms beyond maturity dates to prevent owners from defaulting (i.e., "extend and pretend"). Alby Gallun, "Bubble-era cash-out refinancings pay off for real estate investors," *Crain's Chicago Business*, September 20, 2010.

76. Kris Hudson and A. D. Pruitt, "'Jingle Mail': Developers are giving up on properties," *Wall Street Journal*, August 25, 2010.

77. Quoted in Dan Friedman, "Chicago market profile," Shorenstein Realty Services, August 18, 2000, http://shorenstein.com/about/press/press-article?article_id=2481.

78. D. Marie Steshetz, "From B+ to A," and "Extreme makeovers," *Crain's Chicago Business*, December 5, 2005.

79. Ari Glass, "Historic Building Rehabs: Opportunities, Challenges, and Lessons Learned" (Urban Land Institute presentation, Chicago, May 22, 2013).

80. In addition to pressure from West Loop developers and owners, the lack of action on the 2003 *Central Area Plan* was also seen as motivation for this TIF district designation. Greg Hinz, "City rewriting plan for Loop," *Crain's Chicago Business*, May 29, 2006.

81. City of Chicago, *LaSalle Central Redevelopment Plan* (Chicago: Department of Housing and Economic Development, 2006), https://www.cityofchicago.org/dam/city/depts/dcd/tif/plans/T_147_LaSalleCentralRDP.pdf.

82. Ibid., 27 and 29.

83. The report concluded that 48 of the 93 buildings (53 percent of the total) and 22 of the 42 blocks (52 percent of the total) met these criteria in the proposed area.

84. Twenty-one of the projects converted between 1996 and 2008 became condos, while three became rental apartments.

85. Planner A.

86. Author's calculation based on CoStar Property Analytics data, 1998–2006.

87. Robert Beauregard, "The textures of property markets: Downtown housing and office conversions in New York City," *Urban Studies* 42, no. 13 (2005): 2431–45.

88. Planner A. Buildings that were located in historic districts such as the Michigan Avenue Landmark District, but were not individually landmarked themselves, were then eligible for Class L property tax abatements. Class L property is assessed at 10 percent of its market value (instead of 38 percent) for the first ten years, 15 percent in year eleven, and 20 percent in year twelve, provided that owners invest at least half of the value of the building in an approved rehabilitation project.

89. In 2009 the city passed an ordinance that allowed smaller historic buildings to override the zoning code and qualify for Planned Development status—with no restrictions on changing their use. Offering his support to the ordinance, East Loop alderman Brendan Reilly stated that he had "growing concerns over older building stock in the ward facing high vacancy rates *due to their obsolescence* . . . The city cannot afford for these important buildings to ultimately be abandoned and left to deteriorate because their zoning does not allow for the flexibility necessary for redevelopment" [italics mine]. Steven Dahlman, "'Historic status' ordinance passes city zoning committee," *Loop North News*, September 2, 2009, http://www.marinacityonline.com/news/reilly0902.htm.

90. Other examples include the 1928 headquarters of the Mather Stock Car Company, which was converted to the River Hotel. It received TIF funding to renovate the cupola and stabilize the exterior of the building.

91. Mirabai Auer, "Managing the changing internal landscape: An analysis of office to residential conversions in Chicago's Loop (1995–2006)" (master's thesis, University of Illinois at Chicago, Fall 2008).

92. Steve Daniels, "Tight squeeze: Housing boom displaces small firms, nonprofits," *Crain's Chicago Business*, April 3, 2000.

93. Planner C. However, see series on the city's lapses, including Blair Kamin and Patrick Reardon, "Part 1: The threat to neighborhoods" *Chicago Tribune*, January 13, 2003.

94. The YWCA Building (1929) on Oak Street was also demolished for condos, and the Hotel Dana (1891) in River North, Chicago's oldest running hotel, was torn down to make way for a boutique hotel bearing the same name. "Green" condos were erected on the site of the Hotel LaSalle parking garage (1918), the country's first multistory garage. The State-Lake Theatre (1917) on State Street was torn down for a mixed-use building that included the studios of the Joffrey Ballet as well as several floors of condominiums and retail space. Thanks to the Chicago Department of Buildings for data on demolition permits.

95. Lawrence Downes, "Defacing the skyline, a heartless act in the heart of Chicago," *New York Times*, October 30, 2004.

96. Bradford J. White, "The politics of preservation: Where the public meets the private" (Arts and Humanities in Public Life Conference, University of Chicago Cultural Policy Center, 2003), http://culturalpolicy.uchicago.edu/politics-preservation-where-public-meets-private.

97. Alby Gallun, "Buck plays TIF angle," *Crain's Chicago Business*, August 1, 2005.

98. Pension Fund Advisor A.

Chapter Seven

1. *Schneider v. District of Columbia*, 117 F. Supp. 705 (D.D.C. 1953), 719.

2. Kevin Lynch, *What Time Is This Place?* (Cambridge, MA: MIT Press, 1972; 2001), 39.

3. Paul Knox, "Creating ordinary places: Slow cities in a fast world," *Journal of Urban Design* 10, no. 1 (2005): 3–13; Heike Mayer and Paul Knox, "Slow cities: Sustainable places in a fast world," *Journal of Urban Affairs* 28, no. 4 (2006): 321–34.

4. David Gann, *Building Innovation: Complex Constructs in a Changing World* (London: Thomas Telford Publishing, 2000).

5. Kheir Al-Kodmany and Mir Ali, *The Future of the City: Tall Buildings and Urban Design* (Southampton, UK: WIT Press, 2013).

6. Edward Glaeser, *Triumph of the City* (New York: Penguin, 2011), 148.

7. Ibid., 151. Moreover, New York City's restrictions on housing development are not substantially more onerous than Chicago's. The success of the Bloomberg administration in upzoning significant areas of the city and sanctioning the construction of 170,000 new housing units during the Millennial Boom calls into question Glaeser's claim. See "Reshaping New York," *New York Times*, http://www.nytimes.com/newsgraphics/2013/08/18/reshaping-new-york/.

8. Enrico Moretti, *The New Geography of Jobs* (New York: Mariner Books, 2013).

9. Robert E. Lang, Thomas W. Sanchez, and Asli Ceylan Öner, "Beyond Edge City: Office geography in the new metropolis," *Urban Geography*, 30, no. 7 (2009): 726–55.

10. Appraiser A.

11. In 2010 office vacancy rates in the Chicago suburbs were more than twice those of the downtown market. Eddie Baeb, "'It's a slow drip process to better health,'" *Crain's Chicago Business*, December 27, 2010.

12. Marshall Berman, *All That Is Solid Melts into Air* (New York: Penguin, 1982).

13. In recent years even historians have embraced such arguments. In an editorial in the *New York Times*, Kenneth Jackson lambasted critics of New York City's attempt to upzone the East Midtown section of Manhattan, arguing that the city would not be able to compete with Hong Kong for upwardly mobile residents if it did not allow developers to replace mid-rise apartment buildings with new supersized skyscrapers. Kenneth Jackson, "Gotham's towering ambitions," *New York Times*, August 29, 2013.

14. SamInTheLoop, "Chicago/River Point (444 West Lake)/730 Ft/52 Floors," *Skyscraper Page Forum* (June 20, 2007), http://forum.skyscraperpage.com/showthread.php?t=131220&page=9.

15. Brokers B, C, and D. See also Michael Ball, Colin Lizieri, and Bryan MacGregor, *The Economics of Commercial Property Markets* (London: Routledge, 1998); Neil Dunse, Chris Leishman, and Craig Watkins, "Classifying office submarkets," *Journal of Property Investment & Finance* 19, no. 3 (2001): 236–50; Thomas Brennan, Roger E. Cannaday, and Peter F. Colwell, "Office rent in the Chicago CBD," *Real Estate Economics* 12, no. 3 (1984): 243–60. When oversupply is extreme, however, the chances of upward mobility are greater. See William Hauser, "The economic battle in America's office markets revisited: Dynamics of Class A, B, and C office space in the Chicago market" (ULI Research working paper 633, Urban Land Institute, Washington, DC, 1993).

16. George Nelson, "Obsolescence," *Perspecta* 11 (1967): 170–76.

17. Richard Lloyd, *Neo-Bohemia: Art and Commerce in the Postindustrial City* (New York: Routledge, 2005).

18. Sharon Zukin, *Loft Living: Culture and Capital in Urban Change* (New Brunswick, NJ: Rutgers University Press, 1989).

19. Camilo Jose Vergara, *American Ruins* (New York: Monacelli, 1999).

20. Julia Christenson, *Big Box Reuse* (Cambridge, MA: MIT Press, 2008).

21. John Colapinto, "The real-estate artist: High-concept renewal on the South Side," *New Yorker*, January 20, 2014.

22. Renovation is more mainstream in cities, such as San Francisco, where supply is constrained (because of either topography or planning regulations) and demand, particularly from a younger demographic, is growing. See Michael Kimmelman, "Urban renewal, no bulldozer: San Francisco repurposes old for the future," *New York Times*, June 1, 2014.

23. Christopher Mele, *Selling the Lower East Side: Culture, Real Estate, and Resistance in New York City* (Minneapolis: University of Minnesota Press, 2000); Lloyd, *Neo-Bohemia*; Zukin, *Loft Living*.

24. Jeff Byles, *Rubble: Unearthing the History of Demolition* (New York: Broadway Books, 2006).

25. Franklin and Associates, "Characterization of municipal solid waste in the United States," report no. EPA530-R-98-007 (Washington, DC: Environmental Protection Agency, 1998), http://www.epa.gov/osw/nonhaz/municipal/pubs/msw97rpt.pdf; Michelle Colledge, "Construction and demolition landfills: Emerging public and occupational health issues," *Journal of Environmental Health* 71, no. 2 (2008): 50–52.

26. Most of the debris dumped illegally was generated at North Side construction sites, and after aldermen were bribed, it was dumped on the West and South Sides. Grassroots responses to these injustices helped catalyze a citywide movement for environmental justice and building material reuse. David Naguib Pellow, *Garbage Wars: The Struggle for Environmental Justice in Chicago* (Cambridge, MA: MIT Press, 2004).

27. See, for example, Zhenjun Ma, Paul Cooper, Daniel Daly, and Laia Ledo, "Existing building retrofits: Methodology and state-of-the-art," *Energy and Buildings* 55 (2012): 889–902; National Trust for Historic Preservation, *The Greenest Building: Quantifying the Environmental Value of Building Reuse* (Washington, DC: National Trust for Historic Preservation, 2012).

28. Lynch, *What Time Is This Place?*, 112.

29. Luci Ellis, "The housing meltdown: Why did it happen in the United States?" (working paper 259, Bank for International Settlement, September 2008), http://www.bis.org/publ/work259.htm.

30. Jim Holway, Don Elliott, and Anna Trentadue, *Arrested Developments: Combatting Zombie Subdivisions and Other Excess Entitlements*, Policy Focus Reports (Cambridge, MA: Lincoln Institute of Land Policy, 2014).

31. Lynn Becker, "Are We Dead Yet?," *Repeat*; http://lynnbecker.com/repeat/are%20we%20dead%20yet/arewe3.htm.

32. Real Estate Consultant A and Planner B.

33. Louis A. Galuppo and Charles Tu, "Capital markets and sustainable real estate: What are the perceived risks and barriers?," *Journal of Sustainable Real Estate* 2 (2010): 143–59.

34. Robert Lang and Jennifer LeFurgy, *Boomburbs: The Rise of America's Accidental Cities* (Washington, DC: Brookings Institution, 2007).

35. Reid Ewing, Rolf Pendall, and Don Chen, *Measuring Sprawl and Its Impact* (Washington, DC: Smart Growth America, 2003); Lang, Sanchez, and Öner, "Beyond Edge City." The Chicago region was not immune to the problem of overbuilt suburbs, especially in distant exurbs such

as Kendall County (fifty miles southwest of the Loop). Dirk Johnson, "Briefly the new frontier, exurbs see a bust after boom," *New York Times*, August 20, 2011.

36. Arthur C. Nelson, "The mass market for suburban low-density development is over," *Urban Lawyer* 44, no. 4 (2012): 29–40.

37. Allen Goodman and Thomas Thibodeau, "Where are the speculative bubbles in U.S. housing markets?," *Journal of Housing Economics* 17 (2008): 117–37; Edward Glaeser, Joseph Gyourko, and Albert Saiz, "Housing supply and housing bubbles," *Journal of Urban Economics* 64, no. 2 (2008): 198–217.

38. See Table 1.1. Author's calculation based on CoStar Property Analytics data.

39. William Wheaton and Raymond Torto, "Vacancy rates and the future of office rents," *Journal of the American Real Estate and Urban Economics Association* 16, no. 4 (1988): 430–55; Karl Case, "The central role of home prices in the current financial crisis: How will the market clear?," *Brookings Papers on Economic Activity* (Fall 2008): 161–93; Ko Wang and Yuqing Zhou, "Overbuilding: A game theoretic approach," *Real Estate Economics* 28, no. 3 (2000): 493–522.

40. Brent Ryan and Daniel Campo, "Autopia's end: The decline and fall of Detroit's automotive manufacturing landscape," *Journal of Planning History* 13, no. 1 (2013).

41. David Dowall, "Planners and office overbuilding," *Journal of the American Planning Association* 52, no. 2 (1986): 131–32.

42. Melville Branch, "Don't call it city planning: Misplaced densification in large U.S. cities," *Cities* 3 (November 1986): 290–97; Holway, Elliott, and Trentadue, *Arrested Developments*.

43. Real estate and banking crises tend to be correlated. Richard Herring and Susan Wachter, "Bubbles in real estate markets" (Zell/Lurie Real Estate Center working paper 402, Wharton School, University of Pennsylvania, 2002), http://realestate.wharton.upenn.edu/research/papers/full/402.pdf; E. Philip Davis and Haibin Zhu, "Bank lending and commercial property cycles: Some cross-country evidence," *Journal of International Money and Finance* 30, no. 1 (2011): 1–21; Edward Glaeser, "A nation of gamblers: Real estate speculation and American history" (working paper 18825, National Bureau of Economic Research, 2013), http://www.nber.org/papers/w18825.

44. Stijn Claessens, Ayhan Kose, and Marco Terrones, "The global financial crisis: How similar? How different? How costly?," *Journal of Asian Economics* 21, no. 3 (2010): 247–64.

45. Ellis, "Housing meltdown."

46. Brian Lewis, "U.S. commercial property prices decline on distressed sales, Moody's says," *Bloomberg News*, June 22, 2011.

47. Jane D'Arista and Korkut Ertürk, "The monetary explanation of the crisis and the ongoing threat to the global economy," *Challenge*, March–April 2010, 5–29.

48. It constituted roughly a third of total assets held by the nonfinancial private sector. Edward Leamer, "Housing IS the business cycle" (working paper 13428, National Bureau of Economic Research, 2007), http://www.nber.org/papers/w13428.

49. Christopher Crowe, Giovanni Dell'Ariccia, and Deniz Igan, "How to deal with real estate booms: Lessons from country experiences," *Journal of Financial Stability* 9, no. 3 (2013): 300–319.

50. Richard Stanton and Nancy Wallace, "CMBS subordination, ratings inflation, and regulatory-capital arbitrage" (working paper, Fisher Center for Real Estate and Urban Economics, University of California, Berkeley, 2012).

51. Crowe, Dell'Ariccia, and Igan, "How to deal with real estate booms."

52. Max Page, *The Creative Destruction of Manhattan, 1900–1940* (Chicago: University of Chicago Press, 1999).

53. John Coffee, "The political economy of Dodd-Frank: Why financial reform tends to be frustrated and systemic risk perpetuated," *Cornell Law Review* 97, no. 5 (2012): 1020–82; Sandra Suarez and Robin Kolodny, "Paving the road to "Too Big to Fail": Business interests and the politics of financial deregulation in the United States," *Politics and Society* 39 (2011): 74–102.

54. Dowall, "Planners and office overbuilding," 132.

55. Arthur C. Nelson, Rolf Pendall, Casey J. Dawkins, and Gerrit J. Knaap, *The Link Between Growth Management and Affordable Housing: The Academic Evidence* (Washington DC: Brookings Institution, 2004), http://www.brookings.edu/research/reports/2002/02/housing affordability.

56. Paul Lewis and Max Neiman, *Custodians of Place: Governing the Growth and Development of Cities* (Washington, DC: Georgetown University Press, 2009).

57. While fees have relatively weak effects on new construction, regulations that lengthen the development process or directly constrain new development have stronger effects. Christopher Mayer and C. Tsuriel Somerville, "Land use regulation and new construction," *Regional Science and Urban Economics* 30 (2000): 639–62; John Landis, "Do growth controls work? A new assessment," *Journal of the American Planning Association* 58 (1992): 489–508.

58. In some places, particularly those that could grow at a rapid pace but elect not to, preserving the way of life of incumbent residents is an underlying motivation.

59. Rolf Pendall, "Local land use regulation and the chain of exclusion," *Journal of the American Planning Association* 66, no. 2 (2000): 125–42; Gerrit J. Knaap and Arthur C. Nelson, *The Regulated Landscape: Lessons of Statewide Planning from Oregon* (Cambridge, MA: Lincoln Institute of Land Policy, 1993).

60. Julian Juergensmeyer and James C. Nicholas, "Loving growth management in the time of recession," *Urban Lawyer* 417, no. 1 (Fall 2010/Winter 2011): 4–43, 42.

61. Leamer, "Housing IS the business cycle."

62. Planner B. See similar comments from city administrators in Susan Fainstein, *The City Builders: Property, Politics, and Planning in London and New York, 1980–2000,* 2nd ed. (Lawrence: University Press of Kansas, 2001), 73.

63. States such as Hawaii and Oregon already require local governments to adopt growth management and oversee local development activities stringently.

64. Douglas Noonan, "Finding an impact of preservation policies: Price effects of historic landmarks on attached homes in Chicago, 1990–1999," *Economic Development Quarterly* 21, no. 1 (2007): 17–33.

65. Each $1 million spent on nonresidential historic rehabilitation created twice as many new jobs as new construction as well as more in building income, taxes, and wealth. David Listokin and Michael L. Lahr, *Economic Impacts of Historic Preservation* (New Jersey Historic Trust, 1997). See also *Older, Smaller, Better: Measuring How The Character of Buildings and Blocks Influences Urban Vitality* (Washington, DC: National Trust for Historic Preservation, 2014).

66. Ståle Navrud and Richard C. Ready, eds., *Valuing Cultural Heritage: Applying Environmental Valuation Techniques to Historic Buildings, Monuments and Artifacts* (London: Edward Elgar, 2002).

67. Lynch, *What Time Is This Place?*

68. Taking on the costs of historic renovations is considerably less attractive when there is little hope of finding occupants for the refurbished buildings, as in the case of abandoned automobile plants in Detroit. Ryan and Campo, "Autopia's end."

69. Giorgio Cavaglieri, "Plus factors of old buildings," in *Economic Benefits of Preserving Old Buildings* (Washington, DC: National Trust for Historic Preservation, 1976), 55–56.

70. Rehabilitation of a similarly sized building can lead to cost savings of 3 to 16 percent compared with new construction (if demolition is involved). Donovan Rypkema, *The Investor Looks at an Historic Building* (Washington, DC: National Trust for Historic Preservation, 1991). Building energy consumption per square foot for offices over eighty years old was almost 40 percent lower than for buildings built after 1990. Constantine Kontokosta, "Predicting building energy efficiency using New York City benchmarking data" (presented at the ACEEE Summer Study on Energy Efficiency in Buildings, New York University, 2012), http://aceee.org/files/proceedings/2012/data/papers/0193-000114.pdf.

71. Randall Mason, *Economics and Historic Preservation: A Guide and Review of the Literature* (Washington, DC: Brookings Institution, 2005).

72. Ari Glass, "Historic Building Rehabs: Opportunities, Challenges, and Lessons Learned" (Urban Land Institute presentation, Chicago, May 22, 2013).

73. Due to its decades of occupancy by the same company, the brokerage community was unaware of the Wrigley Building's assets, as few had ever stepped inside. Robert Sharoff, "Modern makeover for Wrigley Building, long a hallmark of Chicago's skyline," *New York Times*, December 10, 2013.

74. Lynch, *What Time Is This Place?*, 112. The counterargument is that it would be unusual for a corporation to compensate a company whose products they replaced (recall Nelson's example of the vacuum cleaner and the broom, above). In contrast to the case of a new consumer product making an older one obsolete, however, the public nature of the built environment means that the impact of the older product's diminished utility is more widely felt.

75. Crowe, Dell'Ariccia, and Igan, "How to deal with real estate booms."

76. Keith Bradsher, "Government policies cool China's real estate boom," *New York Times*, November 10, 2011.

77. Other provisions that should influence secondary markets include the Volcker rule, which prohibits large banks from proprietary trading and from investing in hedge funds and private equity funds. This rule, however, does not apply to many of the medium-sized regional banks that are highly exposed to commercial real estate loans.

78. Coffee, "Political economy of Dodd-Frank."

79. Mary L. Schapiro, "Testimony on Implementation of the Dodd-Frank Wall Street Reform and Consumer Protection Act by the U.S. Securities and Exchange Commission," Senate Committee on Banking, Housing, and Urban Affairs (September 30, 2010), http://www.sec.gov/news/testimony/2010/ts093010mls.htm.

80. Karl Case, "Taxes and speculative behavior in land and real estate markets," *Review of Urban & Regional Development Studies* 4, no. 2 (1992): 226–39.

81. Ping Cheng, Zhenguo Lin, and Yingchun Liu, "Illiquidity, transaction cost, and optimal holding period for real estate: Theory and application," *Journal of Housing Economics* 19 (2010): 109–18.

82. Todd Sinai and Joseph Gyourko, "The asset price incidence of capital gains taxes: Evidence from the Taxpayer Relief Act of 1997 and publicly-traded real estate firms," *Journal of Public Economics* 88, no. 7 (2004): 1543–65.

83. The few studies to analyze transaction taxes have found mixed effects on property prices. See Case, "Taxes and speculative behavior."

84. Raymond Owens, "Commercial real estate overbuilding in the 1980s: Beyond the hog cycle" (working paper 94-06, Federal Reserve Bank of Richmond, April 22, 1994), http://www.richmondfed.org/publications/research/working_papers/1994/pdf/wp94-6.pdf.

85. "The planning of mutations necessarily implies the knowledge and forecasting of ob-

solescence patterns." Manfredi Nicoletti, "Obsolescence," *Architectural Review* 143 (1968): 413–15, 413.

86. Lynch, *What Time Is This Place?*, 73. This raises the question of just how far into the future planners should seek to predict. Lynch is intentionally vague here, making a case for the "middle range." He claims that planners often have realistic short-term expectations, but that their visions for the future are thin.

87. Avi Friedman, "Design for change: Flexible planning strategies for the 1990s and beyond," *Journal of Urban Design* 2 (1997): 277–95.

88. Paul Spencer Byard, *The Architecture of Additions: Design and Regulation* (New York: W. W. Norton, 1998).

89. Maria-Christina Georgiadou, Theophilus Hacking, and Peter Guthrie, "A conceptual framework for future-proofing the energy performance of buildings," *Energy Policy* 47 (2012): 145–55.

90. Deconstruction is the term used to describe an alternative to demolition whereby buildings are broken down in such a manner that salvages as many of the reusable materials as possible.

91. Although more modular and flexible buildings are easier to repurpose and deconstruct, there is a risk that they would also be more generic. Older buildings are appealing because they were customized to earlier needs, but this same feature also reduces their second-hand value.

92. Rex LaMore and Michelle LeBlanc, *Planning Policies and Regulations That Can Reduce the Practice of Private Property Abandonment in the United States* (Lansing, MI: Center for Community and Economic Development, 2013).

93. This suggestion echoes a proposal made a century ago by the Building Owners and Managers Association for business owners to set up a fund to replace obsolete office buildings (i.e., once they hit thirty). Earle Shultz, "The effect of obsolescence on the useful and profitable life of office buildings," *Buildings and Building Management*, August 7, 1922, 13–23.

94. Bubbles, crises, and risk are typically "post hoc denunciation[s]," which are visible only in hindsight. Janet Roitman, *Anti-crisis* (Durham, NC: Duke University Press, 2013), 73.

95. Mayer and Somerville, "Land use regulation and new construction"; James Shilling, Clemon Sirmans, and Krisandra Guidry, "The impact of state land use controls on residential land values," *Journal of Regional Science* 31, no. 1 (1991): 83–92; John Quigley and Steven Raphael, "Regulation and the high cost of housing in California," *American Economic Review* 95, no. 2 (2005): 323–28.

96. Arthur Nelson, Raymond Burby, Edward Feser, Casey Dawkins, Emil Malizia, and Roberto Quercia, "Urban containment and central-city revitalization," *Journal of the American Planning Association* 70, no. 4 (2004): 411–25; Nelson, Pendall, Dawkins, and Knaap, "Growth management and affordable housing."

97. Moreover, if the most overbuilt markets are already some of the most expensive ones, growth controls imposed in those same markets might do little to change the distribution of property values and would not make property less affordable.

98. Neil Harris, *Building Lives* (New Haven, CT: Yale University Press, 1999).

99. Rayner Banham, "Vehicles of desire," *Art*, no. 1 (September 1955), 3.

Epilogue

1. Steve Smith, "Corporate Headquarter Relocations in the Downtown Office Market" (Urban Land Institute presentation, Chicago, Union League Club, September 12, 2012).

2. Abe Tekippe and Thomas Corfman, "Starting over," *Crain's Chicago Business*, February 18, 2013.

3. Lender E.

4. Nadja Brandt and John Gittelsohn, "Chicago office tower to be sold for record $850 million," *Bloomberg*, May 20, 2014, http://www.bloomberg.com/news/2014-05-19/irvine-co-agrees-to-buy-chicago-tower-for-850-million.html.

5. Broker D. Commenting on Hines's River Point, fellow developer Steve Fifield noted, "We've had a four-year hiatus [on leasing], and a lot of companies are bursting at the seams because they've deferred real estate decisions. If Chicago goes back to absorbing a million-and-a-half square feet a year, the West Loop can only accommodate the demand for five or six years. The West Loop could literally be filled in." Quoted in "Hines digs in with another downtown office proposal," *Curbed Chicago*, July 11, 2012, http://chicago.curbed.com/archives/2012/07/11/hines-digs-in-with-another-downtown-office-proposal.php.

6. Ryan Ori, "Tenants flee suburban office parks," *Crain's Chicago Business*, October 1, 2012; Lauren Weber, "Companies say goodbye to the 'burbs: Young talent wants to live in Chicago, not Libertyville," *Wall Street Journal*, December 4, 2013.

7. Ryan Ori, "Downtown office vacancy lowest in 4 years" *Crain's Chicago Business*, September 30, 2013.

8. Developer D.

9. Alby Gallun, "Michigan Avenue's dead zone comes back to life," *Crain's Chicago Business*, July 8, 2013.

10. Alby Gallun, "Building bash: More developers plan downtown apartments," *Crain's Chicago Business*, February 6, 2013.

11. *Crain's Chicago Business*, "The next downtown office tower: One and done?," case study, February 25, 2013, http://www.chicagobusiness.com/realestate/20130418/CRED03/130419770/buy-crains-case-study-the-next-downtown-office-tower.

12. Ryan Ori, "Law firm locks in lease at new office tower," *Crain's Chicago Business*, July 9, 2013.

13. Bryce Myers, "Tishman pulls trigger on new Alliance Center office tower in Buckhead," *CoStar News*, July 25, 2014, http://www.costar.com/News/Article/Tishman-Pulls-Trigger-On-New-Alliance-Center-Office-Tower-in-Buckhead/162546.

14. Homer Hoyt, "The effect of cyclical fluctuations upon real estate finance," *Journal of Finance* 2, no. 1 (1947): 51–64. On institutional memory loss, see Allen Berger and Gregory Udell, "The institutional memory hypothesis and the procyclicality of bank lending behavior," *Journal of Financial Intermediation* 13, no. 4 (2004): 458–95. On the absence of learning over time in real estate, see Alan Rabinowitz, *The Real Estate Gamble* (New York: American Management Association, 1980).

Index

Page numbers in italics refer to illustrations.